The Recognizing and Recording Reform in Mathematics Education Project: Insights, Issues, and Implications

edited by

Joan Ferrini-Mundy
and
Thomas Schram

National Council of Teachers of Mathematics

Library of Congress Cataloging-in-Publication Data:

ISBN: 0-87353-433-6

Printed in the United States of America

Table of Contents

Authors

Joan Ferrini-Mundy
Professor of Mathematics
Department of Mathematics
University of New Hampshire
Durham, NH 03824

Beverly Ferrucci
Associate Professor of Mathematics
Department of Mathematics
Keene State College
Keene, NH 03431

Karen Graham
Associate Professor of Mathematics
Department of Mathematics
University of New Hampshire
Durham, NH 03824

Loren Johnson
Visiting Assistant Professor of
  Mathematics Education
Mathematics Department
University of California—Santa
  Barbara
Santa Barbara, CA 93106

Laura Coffin Koch
Associate Professor of Mathematics
General College
University of Minnesota
Minneapolis, MN 55455

Joanna O. Masingila
Assistant Professor of Mathematics
  and Mathematics Education
Department of Mathematics
Syracuse University
Syracuse, NY 13244-1150

Geoffrey Mills
Associate Professor of Education
Department of Education
Southern Oregon State College
Ashland, OR 97520

Thomas Schram
Associate Professor of Education
Department of Education
University of New Hampshire
Durham, NH 03824

Patricia P. Tinto
Assistant Professor of Teaching and
  Leadership
Teaching and Leadership Program
Syracuse University
Syracuse, NY 13244

Acknowledgments

This report was prepared with support from the Exxon Education Foundation, through an award to the National Council of Teachers of Mathematics.

Joan Ferrini-Mundy is on leave, from 1995 through 1998, at the National Research Council. In no way should the content of this report be ascribed to the National Research Council.

We appreciate the openness and assistance of the teachers, students, parents, and administrators in the seventeen R^3M project sites and acknowledge the insightful contributions of the R^3M project documenters. Special thanks are due to Loren Johnson for his steadfast commitment and ongoing assistance to this project.

Abstract

In this monograph, we present findings from the Recognizing and Recording Reform in Mathematics Education (R^3M) project, a study designed to assess the influence of, depth of knowledge about, and interpretation of the National Council of Teachers of Mathematics' *Standards* documents in several school and district sites. In R^3M, a team of researchers studied 17 diverse sites engaged in attempts at significant change in mathematics teaching and learning. We begin by presenting the project's evolution and history, its theoretical and conceptual perspectives, and a discussion of the methodological challenges encountered. We then present 4 case studies from 4 very different project sites. The final chapter summarizes what we learned from the case study sites as well as the 13 other sites, and concludes with a discussion of the implications of the research.

The R^3M project was sponsored by the National Council of Teachers of Mathematics and funded by the Exxon Education Foundation. During the 1992–1994 school years, after a national solicitation for site nominations, formation of the project research team, and development of the methodology to be used, site visits were conducted. In all sites, schools, teachers, administrators, and parents were grappling with the challenges of implementing different approaches to mathematics teaching, introducing new curriculum, or changing programmatic directions. Through the case studies, we portray from the perspective of the sites the stories of mathematics education change in the 4 settings. Features of the different sites include the strong influence of a principal's visionary leadership, the impact of partnerships with industry and university mathematicians, and the effect of reform efforts on the work of an elementary school mathematics specialist.

We also discuss the challenges of this type of research, including finding a balance between identifying "model sites" and more realistically conveying the obstacles and tensions inherent in this type of change, and the balance between being descriptive and being interpretive. In the conclusion, we discuss implications for instructional practice, administrators and policy makers, future research, and the future of mathematics standards.

Introduction

The Recognizing and Recording Reform in Mathematics Education Project

Joan Ferrini-Mundy

The nation's educational communities are engaged in a variety of reform processes, in response to influential reports and the need for improved opportunities for all our children. In particular, the mathematics education community has been at the forefront of promoting substantial change in mathematics teaching and learning. Documents created by the National Council of Teachers of Mathematics (NCTM)—the *Curriculum and Evaluation Standards for School Mathematics* (NCTM, 1989), the *Professional Standards for Teaching Mathematics* (NCTM, 1991), and the *Assessment Standards for School Mathematics* (NCTM, 1995)—were designed to promote a new, shared vision of mathematics teaching and learning. The documents have been widely disseminated and discussed, and anecdotal evidence indicates that teachers of mathematics are seeking ways to enact the ideas contained in the *Standards* documents.

With generous support from the Exxon Education Foundation, NCTM has been engaged in a multiyear project, Recognizing and Recording Reform in Mathematics Education (R^3M), to assess the influence of, depth of knowledge about, and interpretation of the NCTM *Standards* in several communities; to develop detailed and useful descriptions of "sites of reform"—which might be classrooms, school buildings, or even school districts where significant change in mathematics teaching and learning is occurring; and to assemble and disseminate what is learned in forms that are accessible to a variety of audiences, particularly practitioners.

The release of the *Curriculum and Evaluation Standards for School Mathematics* (NCTM, 1989) was the culmination of a long-term consensus-building process whereby all sectors of the mathematics education community had the opportunity to help formulate a vision for the content and pedagogy of school mathematics for the next decade. The *Professional Standards for Teaching Mathematics* (NCTM, 1991) provides guidance for changing mathematics teaching and offers a set of principles for teaching mathematics in ways that support the vision of the curriculum *Standards*. They "furnish guidance to all who are interested in improving teaching" (NCTM, 1991, p. 7). The publication of the *Assessment Standards for School Mathematics* (NCTM, 1995) expands on the assessment alternatives suggested in the *Curriculum and Evaluation Standards*. The combination of the three *Standards* documents, complemented by other efforts by NCTM and other organizations, forms a foundation for unprecedented change in mathematics education.

In September, 1989, the NCTM Task Force on Monitoring the Effects of the *Standards* recommended that NCTM "monitor their own and other activities designed to implement the *Standards* and to monitor (not conduct) a broader program of research and development" (Schoen, Porter, & Gawronski, 1989, p. 27). A planning grant to consider the monitoring process was awarded to NCTM by the Exxon Education Foundation for the 1991 calendar year; a planning task force for the R^{3}M Project was appointed by the NCTM President and first met in January 1991. A proposal was submitted to the Exxon Education Foundation in February 1992 (Ferrini-Mundy, 1992) and was subsequently funded.

The R^{3}M project became the collective work of a team of 22 researchers working with 17 K–12 school sites. Conceptual and theoretical perspectives were developed to guide the project. Methodologies were adapted from ethnographic traditions to fit the project design and framework. Some interesting and ongoing tensions made the project all the more challenging and intriguing: conducting full-blown case studies and visiting sites for only a short time, describing what we saw in place and describing the evolution of the change process, honoring "model" sites and acknowledging dilemmas, and writing for researchers and writing for practitioners.

One of the difficulties faced by the researchers was how to make the data more manageable and still convey the complexity of change within the sites. To reduce the data to a more manageable form, as part of site visit reports, researchers were asked to identify two or three things that stood out about each site's efforts toward bringing about reform, and each of these characteristics became the basis of a separate story or "scenario." A *scenario*, as used here, is a "story" that captures the most compelling feature of a site through a compilation of field notes and interviews, with relatively minimal interpretation by the researchers. A *compelling feature* is not necessarily a strength of a site, but rather a salient feature worth discussing for the benefit of others. The scenarios have served as the basis for the development of other project presentations, reports, and papers. Two cross-case analyses, drawing on the full database, have been completed (Johnson, 1995; Johnson, 1996).

In this monograph we present both the stories of some of the sites and the story of the project itself. Chapter 1 provides an overview of the goals and conceptual framework, as they were developed and agreed to by a team of diversely experienced researchers that included graduate students, mathematics education researchers, mathematics teacher educators, and cultural anthropologists. Chapter 2 is a reflection on the methodology from the perspective of two of the educational anthropologists on the documentation team. In Chapters 3 through 6, four case studies provide stories of mathematics education change in four diverse settings chosen to be case studies because of the particular richness we found in our visits. Four very diverse case studies compose the chapters of this section. Chapter 3 looks at the reform efforts of Deep Brook Elementary School, which were guided by the visionary leadership of its principal. A university partnership served as a catalyst for the reform efforts at Desert View High School, featured in Chapter 4. As depicted in Chapter 5, an industry partnership provided necessary resources that enabled mathematics teachers at East Collins High School to develop a mathematics program "their way." Finally, the impact of reform efforts on an individual mathematics specialist at Armstrong Elementary School was the focus of the case study presented in Chapter 6. In the summary chapter we provide conclusions and implications that incorporate not only what was learned from the case study sites but what was learned from the other 13 sites as well.

We remind the reader that the case studies are snapshots of one point in time, the "ethnographic present." Surely if we could visit these sites today we would see a different view of mathematics education.

All school and individual names are pseudonyms; we have tried to preserve anonymity and to respect confidentiality.

Chapter 1

Goals and Conceptual Framework

Joan Ferrini-Mundy and Karen Graham

PROJECT GOALS AND DIMENSIONS

The R^3M project was designed to portray the complexity of changing mathematics teaching and learning by describing the efforts and accomplishments of school sites as they work to implement their particular visions of what school mathematics should be. The project goals were—

- to measure the breadth and depth of knowledge about the NCTM *Standards* in various communities;
- to develop useful descriptions of teachers, classrooms, and children in settings where significant attempts at change in mathematics education, consistent with the NCTM *Standards*, seem to be underway;
- to describe the effects of this changed practice on classrooms and on children's learning of mathematics, in ways acceptable as evidence by teachers, policy makers, and the public;
- to increase understanding of the circumstances, forces, and situations in which change in the teaching and learning of mathematics occurs;
- to synthesize and disseminate insights and findings about contextual features that promote and hinder change in mathematics teaching and learning as envisioned in the NCTM *Standards*;
- to assist classroom teachers with the process of change in mathematics education by communicating descriptions of efforts to effect change.

The major emphasis of the project was a documentation and synthesis of practice in which a series of interesting sites of reform were identified and looked at closely in order to learn about the process of change and the interpretation of the ideas presented in the NCTM *Standards* documents in diverse contexts.

DOCUMENTATION AS TRANSFORMATIVE RESEARCH

The mathematics education community is involved in providing direction and vision for mathematics reform, in developing a better understanding of the professional development of teachers of mathematics, and in the challenging process of studying and documenting mathematics education reform as it occurs. A number of major projects are concerned with this process of transformative (Silver, 1990) research; that is, studying "what ought to be." Such studies are critical for many reasons, including—

- the need to provide feedback for continuing efforts at change;
- the need to build a set of examples from practice to inform continued efforts and discussion;
- the need to maintain increasingly deep discussion about reform;
- the need to demonstrate to policy-makers the nature and complexity of mathematics reform (Cohen & Ball, 1990a);
- the need to explain and justify mathematics reform to a broader community (Ferrini-Mundy, 1992; Schoen, Porter, & Gawronski, 1989).

All of these needs provide a rationale for a range of efforts at documenting and studying mathematics change in schools, and several efforts are underway. Early in the Standards movement, the NCTM Research Advisory Committee argued for documentive research (Research Advisory Committee, 1988):

> The *Standards* imply dramatic changes in the nature of precollege mathematics education, and they may also suggest a need for what Silver calls a *transformative* research agenda for mathematics education. Instead of dealing solely with the careful study of "what is" happening currently in the teaching and assessment of mathematical problem solving, research on the transformative agenda will need to deal more broadly with "what ought to be." (p. 341)

Research that studies both the process and nature of change and attempts to characterize implementation and reform activity is of a transformative nature. It is important in the mathematics education community to be able to describe the changes that occur in conjunction with the *Standards* documents and related activities and to be able to explain as best we can how these changes occurred. Such information can actually help the reform process as it evolves.

Apple (1992) describes the *Standards* volumes as a slogan system characterized by a "penumbra of vagueness" embraceable by a diverse set of people, but with enough specificity to offer practitioners something. In the R^3M work, we are learning more about the particulars the documents actually do and do not offer practitioners. Apple's contention, that the documents and the organization need to go further into the policy arena if change is truly to occur, seems well founded. He warns that "without a more thorough grasp of the connections between schooling and these larger power relations, mathematics educators may not have the intellectual resources necessary to make the changes they so clearly urge on us" (p. 429). Remaining attentive to the questions of the policy community and offering contributions from the mathematics education point of view is important.

Researchers and practitioners have called for useful descriptions of practice. Boyer (1990) contends: "It is not enough to suggest active learning and cooperative practices without greater clarity about how teachers might move constructively in these descriptions ... there is need for good description of practice that moves in the direction of the reforms." Along the same line, Shulman (1987) argues for codification of the "practical pedagogical wisdom of able teachers..." and notes that the research community needs to work with the community of practice. This line of thinking has influenced R³M.

Schoen et al. (1989), in suggesting a model for monitoring the effects of the *Standards*, argued that "data provided by monitoring activities would be used to coordinate and guide needed adjustments in other components of the model, including adjustments in the content of the *Standards* and the implementation of that content." This suggests an interpretation of the reform documents as frameworks and guides that are subject to reshaping, revision, and refinement. We hope R³M efforts will influence these "adjustments" in the reform process.

PERSPECTIVES ON CHANGE AND REFORM

Sarason (1991) indicated the dangers to reform that may be imposed by those who are not informed about the complexity of the school system, and he cautioned that being part of the system is no guarantee that one understands the system in any comprehensive way. The "changer" must know the context in which intervention is to take place. Recognizing and recording this context for reform is an important focus of the R³M project.

Sarason further cautioned that "to confuse change with progress is to confuse means with ends." Effecting change is a complex process that those committed to change may not fully realize (Fullan, 1991). In fact, Fullan pointed out, there is often an inverse relationship between "commitment to *what should be changed*" and the "knowledge about *how to work through the process of change*"(p. 95). Both authors argued forcefully that effective reform must not be a piecemeal process but rather should include changing the complex system of the school.

Romberg (1988a) argued that

> The most important barriers to reform are the beliefs, attitudes, and expectations strongly held by all persons involved in education in relation to specific aspects of reform .... The proposed changes are a direct challenge to perceptions held by many persons about the content of mathematics, about what is important for students to learn, about the job of teaching, about what constitutes the work of students, and about the professional roles and responsibilities of teachers and administrators (p. 35).

He also cited as barriers to reform organizational resistance and the possibility of making nominal change by changing labels rather than by changing practices. Seeing reform as "altering a few parts" and overcoming the barriers presented by the cost of reform and the political framework within which schools reside make change difficult.

Finn (1993) was not optimistic that there is a sufficient supply of "terrific math teachers" to bring about the intended change. He asked, "what happens to millions of children whose less-than-gifted instructors rely on prepacked programs, the latest nostrums, and what others tell them is the approved way to proceed?" Finn worried that essential skill building in mathematics will be eliminated if the Standards are implemented. Fullan (1993) contended that our educational system is basically conservative and has a built-in resistance to change. Change is a terribly complex process that requires an enormous amount of human energy, even to make small steps in improving what most people would agree needs to be improved (Fullan, 1993; Sarason, 1991). How do we know what part of this energy will be disruptive and what will be significant to produce change? These are serious concerns that must be confronted if significant change in mathematics learning and teaching is to result.

It is an enormous challenge for teachers to explain and defend their own tentative and uncertain ideas about how to change and improve mathematics teaching and learning to parents who want clear, forceful reassurance that their children are getting the best instruction possible. One teacher spoke of "finding the courage to be the expert." As we are learning from articles in the popular press about the NCTM *Standards*, parental support and parental opposition can both be powerful. Assembling evidence that is convincing for parents and empowering teachers to make their case with confidence are important dimensions of continuing efforts at reform.

INTERPRETATION OR IMPLEMENTATION?

Ball (1992a) provided a helpful discussion of standards, informed by her own participation in the development of the NCTM *Professional Standards for Teaching Mathematics*. She observed that "the standards are intended to direct, but not determine, practice; to guide, but not prescribe, teaching" (p. 34). Tensions follow from such a view of standards: the competing need for both consensus and change, direction and discretion, and guidelines and autonomy. Bybee (1993) considered other relevant paradoxes of leadership, including initiating change and maintaining continuity, and encouraging innovation and sustaining tradition. Another tension that would have fit well in this list is the tension between what is known and what is unknown. The *Standards* documents attempt to propel teachers toward an unfamiliar and somewhat invisible version of practice; in a sense, our energies are being poured into aiming at moving targets. Thus the need for instantiations and interpretations of the Standards is critical.

Ball (1992a) recognized the need to abandon an "instrumental" view about translating the ideas of the *Standards* into classroom practice. There is no algorithm within which change will occur and no guaranteed end result. As a result, multiple interpretations of the *Standards* will occur within classrooms and schools. Porter (1989) advised that the best one should expect of standards is a "context of direction" for change. R^3M takes the position that the multiple interpretations are of great significance because a school site's interpretation could have a profound influence on its personal reform process. These multiple interpretations will generate different implementations of the *Standards* that will be valuable to other schools as they pursue reform.

Several interpretations of the *Standards* are suggested by the literature. First, many teachers seem to hold a notion that the Standards are something to be implemented. This interpretation could give rise to superficial implementation efforts. For example, Reys (1992) described a Parent-Teachers Association Meeting at his child's school at which a parent asked the principal if the school was working with the *Standards*. The principal replied, "Yes, we did those last fall." "Doing them" meant, it turned out, devoting 20 minutes of a staff meeting to discussing *Curriculum and Evaluation Standards for School Mathematics* and distributing copies to all the teachers. Other examples of superficial implementation might include instituting the use of calculators, only to check answers, not to explore new concepts; arranging students in small groups, only to complete worksheets; and using manipulatives in rote ways. R³M efforts are attentive to the possibility that deep change may be preceded or, in some cases, short circuited by the institution of visible but nonsubstantive shifts in practice, and that initial shifts may be shallow by necessity.

Second, there is the notion that the *Standards* serve as a validation of what a teacher, school, or district is already doing. This might be particularly true in sites that were involved in mathematics education reform prior to the release of the *Standards* documents. It would seem that this interpretation has the potential for preventing an ongoing and critical examination of what is being done and how it compares to the recommendations of the *Standards* documents. R³M is concerned with developing a better understanding of this use of the *Standards*.

This project emphasizes that the vision of the *Standards* will not become reality quickly, easily, or without experimentation and false starts. In addition, diverse interpretations and enactments of the *Standards* in the schools create the need for a collection of examples to appropriately portray the reform process. It is essential to learn from the process of implementation and change and to disseminate and share that knowledge openly. It is recognized that the communities, schools, teachers, and classes we visit are in a process of making sense of the *Standards*, according to their contextually based interpretations.

GUIDING PERSPECTIVES IN DOCUMENTATION

We struggled with the issue of how to look at the sites we would visit. Should we develop a checklist of Standards-like indicators and search for their occurrence? Should we try to assign a "Standards implementation score" to our sites? For example, would we be in a position to say that a certain second grade has "implemented 60%" of the Standards? These lines of thinking seemed fruitless and completely inappropriate to the task of understanding the way the *Standards* documents are being interpreted.

Through discussion with the original task force, the project Advisory Board, and the documentation team, we came to the view that the perspective guiding this project should be consistent with the philosophical intentions of the NCTM *Standards*, which seem to be based in constructivist assumptions. We recognized that the communities, schools, classes, and teachers that we would visit were in a process of making sense of the *Standards*, or of more general reform discussion in mathematics, and that this sense-making process would be visible through their practice. We agreed that a primary principle of our philosophical orientation would be the notion of deepening our understanding of the site from the perspective of those involved in the setting. This is consistent with educational anthropology's concern for cultural interpretation. Lightfoot-Lawrence's discussion of portraits (Lightfoot, 1983) proved especially helpful in defining our guiding perspective. Our project shared some design similarities and constraints with her project, *The Good High School*: the sites would be visited by teams; the site visits would be brief; the budget for the project would be modest. She commented

> we were not doing the carefully documented, longitudinal work of ethnographers, although we were interested in many of the same qualitative and interpretive phenomena. We were not creating holistic case studies that would capture multidimensional contexts and intersecting processes, although we wanted to describe schools as cultural organizations and uncover the implicit values that guided their structures and decision making. (pp. 12, 13)

In addition, the need for contextual description, which is central to ethnographic work, was agreed to be important here. Lightfoot-Lawrence's reminder of the "liability common to social scientists: the tendency to focus on what is wrong rather than search for what is right, to describe pathology rather than health ... the uncovering of malignancies and the search for their cures" (p. 10) was helpful. Our intention was, then, to attempt to understand the stories of the sites from the point of view of those within the site. Thus our methodology includes the substantive involvement of a site liaison.

A STUDY, NOT A CONTEST

This type of work is very new for an organization such as NCTM. We regard this project as evolutionary, as one in which we can learn a great deal and one in which we will surely make mistakes and false steps. Consider the difficulty of the following issues: NCTM is the organization responsible for the *Standards;* under the NCTM banner, the R^3M project conducted visits to schools to attempt to understand how the *Standards*, and reform in mathematics more generally, are being interpreted. The press for seals of approval, for recognition and kudos, and for evaluative comments from the "NCTM experts" was overwhelming. We tried to develop language to communicate our intentions clearly with all concerned and avoid the misconception that we had come to sites to judge their application of the Standards, but the mentality in many sites was one of wanting to be the winner and of wanting reassurance that they were "doing it right." An administrator in one site asked, "So, tell me, is our mathematics program really avant garde?" Documenters in another site were greeted with a newspaper release that proclaimed the site was being recognized by NCTM, giving a list of reasons for the recognition, *and* claiming "to be considered for the prestigious designation, a secondary school mathematics department must have succeeded in revising their curriculum in a manner consistent with the current, updated NCTM *Standards*."

METHODOLOGY

The following subsections provide an overview of various phases of the project and the methodology employed.

Baseline Quantitative Data

NCTM contracted Horizon Research, Inc. (HRI) in October 1991 to conduct a pilot study of the implementation of the vision of the *Standards* documents. That study, *The Road to Reform in Mathematics Education: How Far Have We Traveled?* (Weiss, 1992), surveyed teachers of mathematics in grades K–12 from 121 schools in eleven states about their attitudes toward teaching, their instructional practices, and their knowledge of the *Standards*. In addition, HRI conducted telephone interviews with elementary and secondary teachers, college and university faculty, consultants, district curriculum specialists, and directors of mathematics-education projects. The HRI study was designed to be exploratory and was conducted in a short period of time with very limited resources. It did not use the statistical-sampling and data-collection follow-up techniques of more rigorous studies. Nevertheless, the sample seems broadly representative of mathematics teachers in the United States. It provided benchmark data for the more in-depth studies that were to follow.

This study indicated that awareness of the first two *Standards* documents was beginning to develop among the nation's mathematics teaching force, though the more recently published *Professional Teaching Standards* was less well known. Teachers in the sample were generally supportive of many principles consistent with those expressed in the two documents, including the beliefs that manipulative materials helped students understand mathematics, that applications are an appropriate component of mathematics instruction, and that virtually *all* students can learn to think mathematically. Teachers expressed concern about the complexities of alternative assessment strategies and using computers as an integral part of mathematics instruction. Teachers also indicated that many of the topics recommended for study in grades K–12 were not yet an integral part of the curriculum. Traditional forms of instruction, for example having students solve textbook problems, still appeared to predominate. The survey supported the need for staff development activities to help teachers interpret the recommendations of NCTM's *Standards* documents.

These early findings were confirmed by the more recent 1993 National Survey of Science and Mathematics Education, also conducted by HRI. This survey was a carefully constructed study that utilized sound statistical sampling procedures to provide current information about science and mathematics education and to identify trends in teacher background and experience, curriculum and instruction, and the availability and use of instructional resources (Weiss, Matti, & Smith, 1994).

Other data from this survey provided insights into the mathematical and pedagogical preparation of teachers. Secondary teachers generally had the strongest mathematical preparation, but elementary teachers were most willing and best trained to use the instructional strategies recommended in the *Standards* documents. The survey also found that teachers were participating in an increasing number of mathematics in-service activities, but there was little opportunity in schools for teachers to collaborate with each other in developing mathematics curriculum and instruction. Providing quality mathematics education for all students is far from a reality, but there is increasing evidence that less tracking is occurring in mathematics programs.

Qualitative Component

The primary methodologies for the study of the school sites are ethnographic. These methodologies are the most promising for providing data that can be useful and compelling for the public, for practitioners, for policy makers, and for meeting the goals of the R^{3}M project. "Case study research offers a surrogate experience and invites the reader to underwrite the account, by appealing to his tacit knowledge of human situations. The truths contained in a successful case study report, like those in the literature, are guaranteed by the 'shock of recognition'" (Adelman, Jenkins, & Kemmis, 1976). The data-collection phase consisted of two principal stages, *site selection* and *site visits.*

Site Selection

To identify sites where mathematics teaching and learning goals and practices are believed to be consistent with the NCTM Standards, we requested practitioners, researchers, and policy makers in mathematics education to nominate school sites. From the 350 letters mailed, we received 190 recommendations for school sites where the reform efforts toward implementing the NCTM Standards were perceived to be in progress. We contacted each nominated site by letter for information, and received 76 completed "Preliminary Information Questionnaires." Five documenters reviewed the questionnaires and of those selected, chose 26 to contact. Three of the documenters conducted telephone interviews with the 26 sites. From those interviews and the "Preliminary Information Questionnaires," Advisory Board members made the final selection of the 12 sites that composed the first round of our documentation effort. Ultimately, the major criterion in this process was the group's subjective sense of the richness of the site. We asked ourselves "whether we would learn something" there.

Our nominations process failed to yield certain types of schools (e.g., urban schools with high minority populations, schools trying to address the algebra-for-all recommendation of the *Standards*). Our announcement in the *NCTM News Bulletin* (NCTM, 1992a) that invited interested schools in these categories to apply for consideration was unsuccessful. We decided to make direct contacts with mathematics supervisors in an effort to identify additional sites that would provide more diversity in our sample. During the 1993–94 school year, five additional sites were added to the study and four of the original sites were visited for a second time by the documenters.

Site Visits

Initial site visits were conducted from November 1992 through early April 1994. A team of two documenters visited each of the 12 sites for two consecutive days, observing mathematics classes and conducting interviews with teachers, administrators, and other key people involved in the mathematics program. On-site liaisons were designated prior to the visits. In most instances the liaison was a classroom teacher, but occasionally a mathematics supervisor or other administrator served in that capacity. The liaison assisted in the previsit scheduling of interviews and observations, worked out local travel and lodging arrangements, and sent requested materials to the documenters. The preliminary reports prepared by each documentation team were shared with the liaisons for the purpose of getting the site personnel's reactions. It was understood that these reactions would be considered during the writing of the documenters' final report.

During the site visit, the documenters took detailed field notes and audiotaped interviews with site personnel—typically teachers, administrators, mathematics supervisors and community members. The documenters prepared follow-up summaries based on field notes and interviews, and organized around a set of guiding issues discussed in the next section. Documentation reports also included artifacts provided by the site, such as student and faculty handbooks, curriculum guides, lesson materials (worksheets, students' work, teacher lesson plans, etc.), and assessment instruments. The scenarios mentioned earlier also were produced.

Seventeen documenters were involved in the first round of site visits. Ten of these documenters were experienced field researchers, among them, three ethnographers. This core of researchers was complemented by classroom teachers, mathematics supervisors, and graduate assistants. At least one documenter in each site team was an experienced field researcher, and at least one had previous classroom teaching experience in mathematics.

As indicated earlier, four of the original 12 sites visited during the first round of documentation were selected for an additional 4-day visit during the spring and fall of the 1993–94 academic year. The purpose of the second visits was to gather additional data and gain a greater depth of understanding of the sites' mathematics programs. The schedule for the visits was developed by the site liaison, in consultation with the documentation team. Second visits included student interviews, multiple observations of several teachers' mathematics classes, pre- and postobservation teacher interviews, and interviews with administrators, school board members, and parents. Whenever possible, classroom observations were conducted in the same class on two consecutive days. Each of the selected sites had a different focus for its reform efforts and provided perspectives on reform based on what was valued in its mathematics education community.

"Planning for reform," for example, ranged from long-term, extensive preparation by an entire community to an evolving effort in which no preplanning took place. One of the sites was in the early stages of change, whereas another site had been actively engaged in the reform of teaching and learning mathematics for eight years. The nature of shared commitment to *Standards*-based mathematics education varied among the sites. In some cases, entire school staffs seemed to have developed consensus; in others, individual teachers sustained the effort virtually single-handedly. In the four case studies presented here, we saw considerable evidence of shared belief in the value of the reforms.

Orienting Themes

A set of five broad themes guided the documenters' visits. These themes were also used to organize the final write-up for each visit. They also come from the *Standards*, as well as from the lines of thinking described earlier in this section, and enabled us to conduct cross-site analyses.

The documenters were given forms to guide postvisit reports, and were instructed to be mindful of the areas to be addressed in the write-ups. It was in the forms that the orienting themes were made explicit, and that concepts and ideas from the *Standards* were addressed. (See Figure 1.1, on the following page.)

Following the site visits, documenters were asked to develop two or three "scenarios" per site—fairly short pieces that would capture the most compelling features or characteristics of a site. The scenarios were to make heavy use of raw data and to include little explicit author interpretation, in accordance with our decision that the sites' stories should be based on the data gathered at each site. Documenters were given complete autonomy in determining the scenario topics, another methodological decision consistent with our overall ethnographic orientation. Scenarios were shared with site liaisons for reaction, and changes were negotiated, sometimes among documenters, project staff, and site staff. The collection of scenarios served as a basis for other project reporting, and is considered part of the site visit write-up database. In sites where we conducted a second visit, the documenters also developed the more detailed case studies, sometimes drawing on the scenarios. These detailed case studies are presented in subsequent chapters of this monograph.

The following excerpts from project forms convey the nature of the orienting themes:

1. Describe the "mathematical vision" held by the people in the site. (What are their goals for their mathematics program? What kinds of mathematics learning do they hope their students will experience? What features of the *Standards* emerge as they describe their mathematical vision? What do the people in the site see as worthwhile mathematical tasks, or important mathematical ideas? Is there alignment among teachers and administrators concerning this vision? Does the mathematics program emphasize problem solving, communciation, reasoning, conjecturing, and mathematical connections? Can the classrooms be described as mathematical communities? Why?)

Is the mathematical vision that they hold being brought to life in the classroom? What is happening, mathematically, in this school? Provide as much evidence as possible.

2. Describe the "pedagogical vision" held, relative to mathematics, by the people in the site. (What pedagogical philosophy is articulated in the site? How can you tell? What approaches, strategies, and ways of teaching mathematics are important here? What features of the *Standards* emerge as they describe their pedagogical vision? What do the people in the site believe to be effective pedagogical practices? Is there alignment among teachers and administrators concerning this vision?)

Is the pedagogical vision that they hold being brougt to life in the classroom? What is happening, pedagogically, in this school? Provide as much evidence as possible.

3. Describe how contextual features are influencing, both positively and negatively, the teachers' efforts to change their mathematics practice. (What has happened with in-service programs, with outside consultants, with outside funding, and with the school and district?)[

Figure 1.1. R³M orienting themes

Chapter 2

Cross-Disciplinary Teaming in Research on Mathematics Reform: Evolution of Process and Perspective

Thomas Schram and *Geoffrey Mills*

The broad aim of this chapter is to direct attention to the methodological features of multisite, cross-disciplinary descriptive research. Through a focused examination of how these methodological features took shape in the R³M project, we describe our research as an evolution of *process* and of *perspective*. The term *evolution of process* refers to the dynamic, dialectical relationship between the researchers' conceptual frameworks and their definition of meaningful units of analysis. The term *evolution of perspective* refers to our efforts to construct a shared vision of what we were "about" as a research team in terms of our relationships with each other and with those whose experiences we studied. These efforts entailed comparable amounts of compromise and consensus-building.

In setting for ourselves a descriptive, rather than prescriptive, stance, we established a qualitative emphasis in our inquiry that stressed the need for holism, contextualization, emergent analytical categories, and deferred judgment. The following section summarizes our methodological stance and provides a conceptual framework for our discussion of process and perspective.

UNDERSTANDING COMPLEXITY: THE CHALLENGE AND PROMISE OF QUALITATIVE INQUIRY

Attempts to understand and describe the complexity of the commonplace are certainly not exclusive to qualitative inquiry. Nor is any single qualitative strategy among the set of related approaches—interpretive, constructivist, phenomenological, ethnographic, and the like—necessarily better suited than another for grasping the complex world of lived experience from the point of view of those who live it. The appeal of qualitative approaches is commonly attributed to the acknowledgment of multiple or partial truths, the need for contextual and holistic descriptions, and concern for the specific structure of occurrences rather than their general character and overall distribution (Clifford, 1986; Erickson, 1986; Schwandt, 1994; Wolcott, 1990a). Erickson (1986) suggested that qualitative fieldwork is especially appropriate for answering the following questions:

- What is happening, specifically, in social action that takes place in this particular setting?

- What do these actions mean to the actors involved in them, at the moment the actions take place?
- How is what is happening in the setting as a whole (e.g., the classroom) related to happenings at other levels of the system outside and inside the setting (e.g., the school building, a student's family, federal mandates)?
- How do the ways everyday life is organized in this setting compare with the ways social life is organized in a wide range of settings in other places and at other times?

In other words, those who pursue qualitative inquiry deny neither the usefulness nor the necessity of discovering regularities and making generalizations. However, qualitative researchers "are attracted more to a form of investigation that, by considering the extraordinary variability of things, is replete with—and does not shrink from exploring—ambiguity" (Peshkin, 1988, p. 418).

The aim of attending to the complexity and situated details of everyday events can be accomplished through a variety of methods. But as Wolcott (1992) and Erickson (1986) have noted, methods themselves are perhaps the most unremarkable aspect of qualitative work. Participant and nonparticipant observation, informant interviewing, and archival research are what might be termed, more modestly, techniques of watching, asking, and examining (see Wolcott, 1992). Correspondingly, the use of continuous narrative description as a technique—sometimes labeled "thick description" (Geertz, 1973) or what some members of our research team bluntly termed "writing like crazy"—does not necessarily mean that the research being conducted is fundamentally qualitative or interpretive. What makes such work qualitative or interpretive is a matter of the researcher's intent, orienting questions, analytical framework, and basic validity criteria, rather than of procedure in data collection.

This understanding serves to steer us away from the misleading characterization of qualitative inquiry as radically or exclusively inductive. While it is generally true that the specific terms of qualitative inquiry change in response to events in the field, the researcher always identifies questions and conceptual issues of interest prior to beginning a study. Some categories for observation or inquiry are thus determined in advance, while others are not—contributing to an approach in which "induction and deduction are in constant dialogue" (Erickson, 1986, p. 121). Wolcott (1988) likewise affirmed the centrality of the researcher's questions and the ways in which they may be construed and reconstrued in response to changes in the research setting:

> Probably the most serious misunderstanding (and biggest disappointment) about qualitative research is the realization that, just as with quantitative approaches, *we bring our questions with us* to the research setting. We do not sit by, passively waiting to see what a setting is going to tell us. That is why we can't respond satisfactorily to the plea, "Just tell me the steps..." and why it is difficult to answer even a seemingly straightforward question like, "What do I need to do as a participant observer?" Except for the most global advice, field techniques cannot be distilled and described independently from the questions guiding the researcher or the nature of

the research setting. Nor can the art of problem posing be isolated from the complex web of personal motives and persuasions, professional strictures, prevailing paradigms, or current preoccupations inherent in each unique study. (pp. 17–18)

Mindful of these considerations, we have chosen to focus in this chapter not only on the specific methods used to develop our description and analysis, but also on the assumptions, motives, and orientations that underlay our study and that had an impact on our ability to function as a research team. Many of the data collection techniques used in this study are characteristic of other efforts at descriptive research, but it is our belief that the most telling methodological insights of the R³M project stem from the cross-disciplinary nature of the research team. We turn first to the area of data collection and analysis, with particular emphasis on the emergence and negotiation of themes over a 2-year period.

THE EVOLUTION OF PROCESS

The "real mystique" of qualitative inquiry, asserted ethnographer Harry Wolcott (1994), lies more in the process of *using* data than of *gathering* data. As experienced by team members in the R³M study, the process by which researchers transformed what they saw and heard into intelligible accounts necessitated a comfort with initial ambiguity and an openness to repeated attempts to define meaningful units of analysis. Questions persisted throughout this process: "How do we balance breadth of analysis with the need for in-depth understanding of specific settings?" "How do we coconstruct meaningful analytical categories?" "To what extent do we go 'beyond' our data in developing an interpretive framework?" We have organized our discussion to reflect how we grappled with such issues as they relate to typical research categories: site selection, data collection, and data analysis.

Site Selection

Site selection for the R³M project was an important component of the methodology. It was not the project's goal to identify exemplary sites for study, but rather to describe sites of mathematics reform where something could be learned about implementation or interpretation of the Standards.

Site selection was problematic in many ways. For example, to what extent can accurate indicators of meaningful reform activities be elicited through telephone interviews and questionnaires? Compounding this issue was the predictable response of some study participants who equated site selection with "winning a contest." On more than one occasion, documenters were greeted at a site by local press, who believed that the school or district had been selected by NCTM as an example of outstanding mathematics teaching. The documenters' presence seemed to validate the site's selection even before the research had been undertaken. This validation phenomenon will be discussed later when we consider the ethical issues of how to report back to sites the findings of the research, and the complicated process of negotiating and renegotiating informed consent.

In order to identify the sites for the study, the R^{3}M documenters sought recommendations from teachers, researchers, and policy makers in mathematics
education throughout the United States and Canada. Ferrini-Mundy and Graham
describe the process as follows in Chapter 1 of this monograph:

> From the 350 letters mailed, we received 190 recommendations for school sites
> where the reform efforts toward implementing the NCTM Standards were perceived
> to be in progress. We contacted each nominated site by letter for information, and
> received 76 completed "Preliminary Information Questionnaires." Five documenters
> reviewed the questionnaires and of those selected, chose 26 to contact. Three of the
> documenters conducted telephone interviews with the 26 sites. From those inter-
> views and the "Preliminary Information Questionnaires," Advisory Board members
> made the final selection of the 12 sites that composed the first round of our docu-
> mentation effort. Ultimately, the major criterion in this process was the group's sub-
> jective sense of the richness of the site. We asked ourselves "whether we would learn
> something" there. (p. 12)

This process was repeated for later stages of the project that necessitated identification of additional sites. On the basis of our early experiences—and the reality of not "seeing on the ground" what was reported by site liaisons who had a
vested interest in promoting their own sites—we continued to refine this selection process. In particular, we expanded the scope of references for sites and used
different sets of reviewers for repeated reviews of the same sites. However, the
complexity of site selection involving multiple stakeholders in reform efforts
made this a difficult task at best, and one in which we were not guaranteed to
include the most representative sites of reform.

Data Collection

The data collection techniques for the study included the use of observations,
interviews, and reviews of written data sources. Documenters were provided
with five broad areas of interest to orient their site visits. These areas were taken
from the *Standards*, as well as from the literature on mathematics education
reform, and were intended to elicit information about the following:

- the mathematical vision held by the people at the site;
- the vision of mathematics pedagogy held by the people at the site;
- the ways that contextual features are influencing, both positively and negatively, the teachers' efforts to change their mathematical practices;
- the ways that the mathematical and pedagogical practices in this school are
  affecting students;
- the evolution of the mathematics program in this school (Ferrini-Mundy &
  Johnson, 1994, p. 192).

On the one hand, these guiding principles provided the documenters with a
shared conceptual framework for undertaking the field work. On the other, as we
shall see, the impact of the cross-disciplinary perspectives on the conduct, analysis, and writing up of the research was significant, and perhaps overwhelming,
for some members of the research teams.

Observations. The R³M documenters conducted classroom observations as the methodological core of their field work. In discussing his "participant-as-observer" role in the study of the elementary school principal, Wolcott (1984) described the role as one in which

> the observer is known to all and is present in the system as a scientific observer, participating by his presence but at the same time usually allowed to do what observers do rather than expected to perform as others perform. (p. 8)

Pelto (1978), Wolcott (1982, 1984) and Spradley (1980) recognized that observers become involved in research sites to varying degrees and that there are advantages and disadvantages to active participation as well as passive observation. In our case, determining the degree to which the documenters participated was somewhat complicated by the fact that most of them have mathematics education backgrounds and, perhaps more significantly, by the fact that they were typically perceived as official representatives of the NCTM. This added pressure to the field work of some of the documenters because teachers and administrators sought validation of their practice from the "experts." Documenters were instructed to be "nonevaluative" and to respond to requests for validation with an explanation of the study's goal—to describe and understand, but not evaluate, reform efforts that seem to be consistent with the *Standards*.

Faced with these pressures, documenters most often accepted the role of "privileged observer," a role that Wolcott (1984) asserted is the most appropriate role for researchers because of the limited opportunities that exist for becoming a full-fledged, active participant in their study sites. Wolcott also pointed out that there are many opportunities for observing and recording events in schools as a "privileged observer" and that such opportunities are limited only by the endurance of the researcher. Given the relatively short duration of the site visits in this study (2–5 days on any given visit), documenters' visits were tightly scheduled with classroom observations and interviews. Prior to the visits, documenters negotiated details of entrée and access to classrooms, assuring that teachers and students in the sites would not be required to change their normal practices, schedules, or behavior in any way.

To help structure classroom observations, documenters were provided with a "Summary of Classroom Observation" form that included the following headings:

- A sketch of the classroom setting, including any mathematical "decorations" in the room
- The number of students in the room by ethnicity and gender
- Description of classroom practice guided by:
    mathematical topic
    instructional approach
    instructional materials
    mathematical tasks featured
    appropriateness of pedagogy

mathematical communication

teacher's ability to meet mathematical goals

• the students' experiences of the class:

student engagement

student competency in mathematical processes

• typical class or special presentation for documenter?

• anything else of significance?

The data on these feedback forms were intended to serve as supplemental summaries of field notes and not as a substitute for them. Documenters were instructed to complete these forms in rough-draft form during a site visit and complete refinements immediately upon their return. We found in working with these forms that they were indeed supplementary to our field notes and served as effective catalysts for discussion among the documenters. In one sense, they served to facilitate early analysis—an interim step between field notes and the writing of scenarios of classroom practice.

Interviews. Prior to on-site visits, a documenter conducted telephone interviews with site-liaison personnel to develop a demographic profile of each site. These profiles, which were used primarily as a data source to inform the site-selection process, were also helpful to the preparation for site visits. The telephone interviews included questions about such things as school size and structure, ethnic breakdown, and the history of a site's experience with the *Standards.*

Formal and informal ethnographic interviews were conducted with central-office personnel, principals, teachers, mathematics supervisors or specialists, and students. All of the formal interviews were tape-recorded (with the consent of the informants) and later transcribed for the documenters. Agar (1980) discussed the informal ethnographic interview as being negotiable and suggested strategies that allow the researchers to have a ready set of questions to ask informants, based, for example, on the "five *w*s and *h*"—who, what, when, where, why, and how. Similarly, Spradley offered a taxonomy of ethnographic questions categorized as "descriptive, structural, and contrast" questions (Spradley, 1979, p. 223).

Even though documenters were provided with interview schedules to follow, we also recommended that they utilize informal ethnographic interviews whenever possible and, in every possible way, maximize their time on site. Although this somewhat eclectic approach to interviewing is most familiar to experienced qualitative researchers, our experience with the cross-disciplinary teams was that the documenters quickly assumed the "researcher as instrument" role, frequently engaging in informal interviews with informants and taking the time to reconstruct the conversation in the form of field notes as soon as possible after the interview. We also found that documenters often included in the formal ethnographic interviews questions influenced by their own conceptual framework of mathematics education reform. For example, a documenter who was particularly interested in teacher reflection as it relates to reform efforts would include questions that addressed that interest.

We believe that the experience of actually *doing* on-site research was the primary training for a majority of the documenters. Not all of the mathematics education researchers had had prior experience or training in qualitative or ethnographic research techniques, so they, for the most part, relied heavily on the guidance of the qualitative researchers on their site-visit teams. Our experience suggests that the majority of the documenters were "quick studies" of the process; that is, they quickly became comfortable with the open-ended nature of the research process and were willing participants in both the formal and informal interviews.

Written sources of data. Documenters were also encouraged to collect materials at each site that would contribute to their understanding of the mathematics reform efforts at that site. Typically, these written sources of data included curriculum guides, policy documents, minutes of meetings, teacher-made tests, examples of student work, lesson plans, and other such materials to which the documenter was made privy.

Data Analysis

Data analysis was ongoing and dialectic during the course of the project. In order to facilitate the process, the entire research team convened periodically to discuss the project's conceptual framework and to identify common themes that were emerging at different sites around the country. During the first round of site visits in October 1992, documenters met in Washington, D.C., to discuss their experiences. This meeting was also attended by members of the research team who had yet to visit their sites. This meeting provided an opportunity for the research team to develop a common understanding of the research process and to clarify the conceptual framework guiding the project. At a second meeting in February 1993, in Washington, D.C., documenters made their first joint attempt at analysis.

The complexity of this process is captured in the following scenario. The 17 documenters arrived in Washington with a somewhat open-ended agenda for the meeting. Each documentation team reported on their site visits in terms of the categories of pedagogical vision and mathematical vision and then offered their preliminary ideas regarding emergent themes at their individual sites.

This activity, while guided by the parameters outlined above, quickly took on a freewheeling atmosphere, with documenters sharing everything and anything they found interesting at their sites. What emerged from this discussion was an initial attempt to analyze the data through the construction of a "monster dog" (Miles & Huberman, 1994) from which the following guiding questions emerged:

1. What has enabled mathematics reform to occur?
2. How did the concern for reform evolve into a focus?
3. What is sustaining/supporting mathematics reform?
4. What are participants' perceptions of mathematics reform/practice?
5. How can classroom practice be characterized?

Although this is not an exhaustive list of the questions the documenters gen-erated during their initial analysis of the data, it is illustrative of the process we undertook to—in basic terms—provide documenters with an opportunity to talk about their data and voice their perceptions of emerging themes. It also illustrates how questions overlapped and why this list of guiding questions quickly became redundant.

At the completion of the first day's sessions, the documenters were asked to reflect on the questions that had emerged and to come back the next day prepared to talk specifically about the mathematical and pedagogical visions uncovered at their sites and describe the three most outstanding features of their sites with regard to the aims of the study. This approach was meant to help them select from the initial group of sites four to six sites that would be revisited for an extended period of time during the coming fall. However, on the second day the researchers quickly ignored the proposed format and focused immediately and enthusiastically on the outstanding characteristics of their sites. This sharing took on a competitive spirit as individual documenters promoted the mathematical and pedagogical practices they observed at their sites in order to justify a return visit. In spite of assurances from the project director that there would be plenty of opportunities for field work at other sites, a handful of the documentation teams appeared unwilling to let go of their "interesting" sites. Other documenters were more than comfortable with the elimination of their sites from the pool of sites to be revisited and some expressed relief that potentially "touchy" ethical issues regarding a site would be avoided by not revisiting it.

The documenters' presentations reinforced the notion that there is always something to learn in the course of "doing" qualitative research. However, given the fiscal constraints of undertaking this particularly time- and labor-intensive kind of on-site research, there were exigent conditions that necessitated some way of determining which sites of reform had the greatest potential for being documented more fully as case-study sites—sites where we could maximize our in-depth understanding of what was happening with regard to the *Standards* and broader reform efforts.

By the end of the second day, the documenters had agreed on possible groupings of sites that would allow the project to continue and at the same time contribute the most to our understanding of mathematics education reform. This "packaging" of sites enabled researchers to justify and defend once again the use of their particu-lar sites while revisiting the questions that had already emerged. As a group, we attempted to develop criteria that could be applied to our group of sites—what was it exactly that we wanted to convey about these sites? Were we most interested in the catalysts of reform, teacher understanding and ownership of the *Standards*, content and pedagogy, or the process of change in general? Given the project's goal to provide a deep understanding of the process of change, did we also want to include case studies of stability, rather than change, over time?

During the June 1993 meeting of documenters the same process of data analysis and cross-site analysis was undertaken following a period in which the documenters had written two or three scenarios highlighting selected aspects of their sites. When the group reconvened there was another attempt to analyze the data to see what themes were emerging. These themes were intended to capture the lessons learned from the project. The following themes were identified at the June 1993 meeting:

1. Children's experiences with the *Standards*;
2. Teachers' experiences with the *Standards*;
3. Collaborative experiences in mathematics reform.

At the third meeting of the documenters in February 1994 the themes were again examined, but this time with several new documenters on board. With over a year between the initial data-collection and writing periods (in many cases), the documenters developed yet another set of themes (see Figure 2.1) that consolidated much of the earlier work.

---

1. Knowing mathematics and knowing teaching
   - How does teachers' knowledge of mathematics seem to be reflected in their practice, or in their attempts at change?
   - How does the teachers' knowledge of mathematics pedgogy seem to be reflected in theirpractice, or their attempts at change?
2. Teachers making meaning of mathematics reform
   - roles professionals play: "the personalities of reform" notions of leadership, interaction with outside forces, the community, etc.
   - teachers in transition: illustrations of teachers along the "reform continuum"
3. Children's experiences with mathematics reform
   - technology/tools versus toys: the illusion of reform, thinking it's over before it's started, early adopters and early leavers
   - mathematics for all
4. Curriculum and frameworks
   - What do teachers do when they are "textbook free?"
   - Evaluation and assessment: How are reform-oriented settings coping with assessment?
5. Reflecting on mathematics reform
   - The NCTM *Standards* as catalyst or validation: What comes first, the *Standards* or reform?
   - Constant climate of change: Changing school cultures for the "homeostasis of change"
   - Interpreting the Standards: Many possibilities

Figure 2.1. Themes that emerged following cross-site analysis

With the timeline for the project near completion, and pressure from various sources to share "products" from the research with the mathematics and education communities, this list came to represent the "final" set of themes to emerge from the data analysis. For many members of the cross-disciplinary teams, this open-ended data analysis process was frustrating. In just coming to terms with qualitative approaches, one must balance time-intensive field work with the expectation of producing "findings" that can be shared in a straightforward fashion with the researcher and practitioner communities.

We move now to consider the human element in this cross-disciplinary process and, in particular, how we came to understand our ability to function as a research team.

THE EVOLUTION OF PERSPECTIVE

To convey the complexity of people's roles and interactions in our study, we now highlight several pivotal issues that shaped our understanding of researchers' perspectives as they evolved through the process of gathering and analyzing data and constructing research narratives. These issues are intent and validity; the accommodation of multiple perspectives; and the politics and ethics of field work, particularly as revealed in our efforts to construct research narratives and negotiate accounts with sites. We address these issues not as a means of conveying our methodology in detail, but rather because of their value in suggesting procedures that may be of use to other researchers considering a cross-disciplinary-team approach to field-based inquiry.

Intent and Validity: Accommodating Multiple Perspectives

Our first meeting as a complete 17-member documenters' group to sort through preliminary data affirmed our earlier expectations that we would have to reject the straightforward notion of symmetric functioning as a research team. The project director's initial decision to pair documenters who represented different conceptual, experiential, and disciplinary backgrounds anticipated the need to follow a more complementary approach to teaming: As tasks arose, group members assumed different roles suited to their individual strengths and disciplinary backgrounds. This worked well during the site-visit phase of the project, for example, when one member of each documentation team was an experienced field researcher and at least one had previous classroom teaching experience in mathematics. This approach seemed to provide comparable measures of efficiency, reciprocity, and professional development to our overall team effort. (Note, for comparison, the discussion of balance and synergy in team research by Liggett, Glesne, Johnston, Hasazi, & Schattman, 1994.)

As we entered into phases of analysis and interpretation, however, our effort to capitalize on these features of the cross-disciplinary team became increasingly problematic. A foremost concern was that the predominantly *descriptive* emphasis of the R³M project served to reinforce some team members' commitment to understanding the enormous complexity of observed phenomena, whereas it frustrated the intention of others to limit or reduce the variables of study. During our February 1993 data-analysis session involving all the project documenters, one particularly heated exchange between two team members over whether and when to make a follow-up visit to a particular site uncovered deep-seated differences in research perspectives. One member clearly tended to look intensively, but seldom more than once, at a phenomenon, embodying in this stance the structure, orderliness, and economy of a preferred research approach. The other favored a discipline-based emphasis on contextual and holistic description that meant looking again and again at a site in order to capture both its ordinariness and the variety of experiences it had to offer.

Through their varied emphases, each of these researchers revealed different underlying assumptions about the educational process in general and their interpretation of the project's aims in particular. For the first researcher, it was appropriate to assume that valid constructs of mathematics skill development (its antecedents, processes, and consequences) exist and can be used to study many different people and settings over a period of time to obtain consistent measures of meaning. Thus crucial features of "successful" instructional programs can be applied to other settings and used to modify instructional treatments. Research questions tend to be derivatives—whether explicitly or implicitly—of the general question "How can mathematics education be improved?"

For the second researcher, it was important to assume that human behavior and learning are responsive to specific contexts and that the perspectives of participants in particular events are crucial to an understanding of those events. What is taught, learned, and experienced most likely varies by group and setting. For this second researcher, a central question was "Why is mathematics teaching and learning occurring in this way in this setting?" (In clarifying these distinctions, we have drawn in part upon Eisenhart's 1988 discussion of contrasts between mathematics education research and the ethnographic research tradition.)

Despite these differences, our research team generally viewed both perspectives as significant and useful attempts to address a basic concern for validity, generally defined as the trustworthiness of inferences drawn from data (Eisenhart & Howe, 1992). Although criteria for methodological adequacy and validity essentially have been "owned" by the positivist tradition, issues of descriptive, interpretive, and theoretical validity are of primary concern to qualitative researchers (see, for example, Altheide & Johnson, 1994; Eisenhart & Howe, 1992; Maxwell, 1992). Following Maxwell (1992), we judge *descriptive validity* on the basis of how well a given inference addresses the question of the factual accuracy of the research data. *Interpretive validity* relates to the meaning of events for participants, which necessarily involves the making of inferences. *Theoretical validity* builds on but extends beyond the first two types of validity by "explicitly [addressing] the theoretical constructions that the researcher brings to, or develops during, the study" (Maxwell, 1992, p. 291).

In our data collection and analysis we addressed issues of descriptive validity by portraying events as much as possible from the points of view of the actors involved. We relied heavily on students' and teachers' own language and tried to make certain that our use of their language was accurate. We addressed issues of interpretive validity in several ways. First, our interviews and survey instruments included questions that sought to reveal teachers' (or students' or administrators') understanding of and thoughts about mathematics and reform activities at each site. Our concern here was that data originate from a variety of sources. Second, we made comparisons among data sources to confirm or modify interpretations, seeking an accurate account of individual and group perspectives. Third, we analyzed data independently and then, with our codocumenters and subsequently in cross-site discussions, made comparisons among interpretations to highlight similarities and differences in understanding. When differences emerged, we returned to the data to obtain additional insights and construct follow-up questions for informants that might lead to reconsideration of earlier analysis and interpretation.

Our primary heuristic tool when dealing with interpretive validity was triangulation, or the use of multiple methods and data sources. Denzin (1978) identified four basic types of triangulation, all of which we employed in our study:

- *data triangulation*—the use of a variety of data sources (e.g., teachers, students, administrators, parents, curriculum guides, lesson plans, student work);
- *investigator triangulation*—the use of at least two different researchers in each site (In our project we also employed an on-site liaison who, in some cases, participated with the documenters in subsequent off-site analysis sessions.);
- *methodological triangulation*—the use of multiple methods (e.g., interviews, observations, document analysis) to study a single problem;
- *theory triangulation*—the use of multiple perspectives (e.g., anthropologists', constructivists', practitioners') to interpret a single set of data.

The challenge we faced with respect to theoretical validity was to establish the value and integrity of the categories we used to explain events and concepts. As detailed in the first section of this chapter, we were guided initially by interview protocols that blended preestablished categories (e.g., informants' notions of pedagogical and mathematical visions) and open-ended questions designed to elicit "insider" perspectives on local events. Then, shifting our mode of inquiry more deliberately to analytic induction (Goetz & LeCompte, 1981), we independently scanned the data, seeking to identify categories of phenomena and relationships among categories (LeCompte & Preissle, 1993).

Through these preliminary categories we sought to capture regularities in the data through what Erickson (1986) termed "repeated trials at understanding recurrent events." As initial categories were tested against the data, some were dropped, others were refined, and new ones emerged and were, in turn, tested. For those categories that remained, descriptors were written, and simple matrices constructed (Miles & Huberman, 1994) that allowed us to compare data from individuals and sites and identify possible relationships among categories.

Politics and Ethics

Far from being a soft option in social scientific research, qualitatively oriented field work represents a complex craft that entails both coping with multiple contexts and constraints and continually dealing with ethical decisions. Unarguably, such complexities suffuse all social scientific research, to a greater or lesser degree. So, too, does all social scientific research share fundamental dilemmas, such as the tension between the protection of informants and the freedom to conduct research and publish findings.

However, much of qualitative field work is dependent on one individual's perception of the field situation at a given point in time; in effect, the researcher operates as his or her own "research instrument." Consequently, elements such as personality and the nature of interactions with the researched become pivotal in one's handling of the political and ethical features of research. By *political features*, following Punch (1994), we mean a spectrum of contexts, from the micropolitics of personal relations to the cultures of research units and the policies of funding agencies. By *ethical features*, we mean the social and moral obligations generated by field work, especially as they affect the purpose and conduct of the research itself.

Basically, people must be informed of your role—who you are and what you want as a researcher. In turn, you as researcher must balance the requisites of gaining access with the expectation of eventual departure and presentation of findings. In what follows, we highlight some of the political and ethical dimensions of the R^3M study by reflecting briefly on two aspects of our experience, constructing the research narrative and negotiating descriptive accounts with sites.

Internal collaboration: Constructing the research narrative. In collaborative research such as the R^3M project, one's social, moral, and professional obligations often reach inward to members of the research team—particularly, in our case, as we sought to construct a coherent rendering of what we had learned. To do so without losing sight of the distinctive contributions each of us had brought to the research process required that we confront fundamental questions regarding interpretation and orientation.

The first of two major questions we faced was whether some activities or events were more worthy of interpretation than others. If we were to decide—as we did in our first group data analysis session—that all events and activities do not have equal merit, how would we determine our focus? To what extent would we let our selection of phenomena for interpretation be determined by (a) the advice of those in a particular documentation site; (b) the conceptual orientations and theoretical preferences of individual researchers; (c) the manner in which we reconciled the competing needs of understanding complexity and shaping policy; or (d) chance, with the hope that we would obtain a representative sampling of perspectives?

A second question was whether some individuals' interpretations were better than others. Correspondingly, were interpretations of different types or levels of

people in our sites—students, teachers, principals, district administrators—better
than those of others? In her commentary on the ethics of interpretation, McGee-
Brown (1994) asserted that because we are considered the primary "research
instrument" in what we do as qualitative field workers, "it is essential that we
reflect daily on our contacts with participants at a site to determine whether we
are trying and succeeding in assessing all interpretations of all participants in
multiple contexts equally" (p. 16).

In retrospect, it seems we were consistent in our efforts to create as broad a
forum for interpretation as we could while, consciously or unconsciously, promoting the voice of the classroom teacher. This emphasis on the classroom
teacher was in keeping with the expressed aim of the project director to have our
findings "speak" to a community of practitioners (Ferrini-Mundy, personal communication), but it also reflected our general commitment to understanding the
complexity of mathematics reform from the point of view of those who experience it on a day-to-day basis.

Among other considerations, this orientation facilitated our coming to terms with
the *ex post facto* influence of the *Standards* on mathematics reform efforts in many
of our sites. Especially among the mathematics educators on our team, there was a
strong tendency to put the *Standards* first and then overlay a particular school's
experiences upon them, when in fact, and from the local teachers' perspectives, the
Standards were only applicable in retrospect, as a validation for their efforts.

Ultimately, we found direction in our implicit agreement to operate less out of a
perceived need to find a theory or perspective of "best fit," and more according to
our willingness to widen the range of acceptable frameworks for written narratives.
We furthermore reaffirmed our commitment to a descriptive, nonevaluative stance.
The relative newness of qualitative field work for some team members facilitated
their ability to "step back" occasionally and view the process from a certain comforting distance. Critical questions could be asked without getting personal (and
upset) about them. With all of this "widening" and "distancing" came the ability to
examine our disciplinary differences more objectively and attain clarity of purpose
in what we chose to interpret and how we did so. In most but not all cases, this
helped to transform the sometimes blunt, vague, or even contradictory feedback we
gave each other's writing from threatening criticisms into constructive suggestions
leading to new levels of insight and analysis.

External confidentiality: Negotiating accounts with sites. Among the many implications for qualitative researchers in the codes of professional conduct and ethics are
the basic concerns of confidentiality and feedback between researcher and
researched. The presentation of findings, which Woods (1986, p. 60) called the
"ultimate ethical test," is necessarily bracketed by various safeguards to protect the
privacy and identity of one's settings and respondents. The researcher bears the
responsibility of handling and interpreting data in a way that (a) will be true to his
or her data; (b) will fairly represent the perspectives of his or her informants; and (c)
will not harm, embarrass, or in any way endanger the position of informants in their
settings (Phtiaka, 1994).

With each of the case studies presented in this monograph, respondent validation of data was handled in several ways. First, participants were asked to sign a form giving permission for data obtained through observations and interviews to be used anonymously. This afforded participants the opportunity to protect themselves by excluding sensitive issues. It should be noted that we proceeded with the basic assumption, conveyed to informants from the start, that the researcher has the ultimate responsibility for deciding what may or may not be included in the write-up of the study. While exercising caution in the handling of information, particularly if there were reasons to believe that informants were somehow vulnerable, researchers regarded censorship from participants as inappropriate to the construction of an accurate and unbiased account of the situation as experienced. This was conveyed in the preliminary informed-consent forms.

To ensure that researchers' perceptions matched those of participants in each setting, informants were sent rough drafts of scenarios and the complete case studies with the invitation to check them and, if necessary, highlight misperceptions or factual errors they felt were inappropriate or too sensitive to be used. Drafts of the Desert View case study, for example, were reviewed by nearly every teacher, student teacher, and administrator in both the high school and university settings who had been interviewed and/or observed. Teachers also discussed the drafts with students in each site who had been interviewed. The researchers incorporated their suggestions and comments into subsequent revisions of the case study.

In an additional step not typically taken in large-scale studies of this sort, representatives from each of the four case-study sites were invited to the February 1994 documenters' meeting to share in data-analysis sessions and to engage in direct give-and-take regarding the presentation of findings from their sites. Their participation enhanced the collaborative nature of the overall project and injected a genuine sense of immediacy and integrity into our interpretive efforts.

The major dilemma we faced in our relationships to the study sites pertained to the participants' perceptions of who or what we represented in conducting this study. Specifically, the link between the R^3M project and the NCTM contributed directly to the sites' perceived need to be seen "in best performance." On the one hand, this was a natural extension of the notion that researchers, beyond their obligation to cause participants the "least possible harm," might actually do them some good! It was also manifested in the periodic requests from teachers and administrators for feedback on what we had observed: "So, was that a good lesson?" "Is that what you were looking for?" "How can I improve my presentation?"

On the other hand, as noted in our discussion of site selection, this study was not a contest to be won or lost. Participants who perceived it as such indirectly and unintentionally threatened to compromise our ability as researchers to capture "useful descriptions" of significant attempts at change—descriptions that necessarily included documentation of struggles as well as successes. We were guided by our research aims to suspend judgments on teachers' performance, for example, by responding to their requests for feedback with a brief explanation of the descriptive emphasis of our study and an assurance that specific or identifiable observations or comments would be embedded in the "big picture" of reform efforts occurring in the setting. This was not entirely satisfactory to them or to us, but it enabled us to deflect, if not entirely avoid, the constraining perception of our project team as an agency that somehow granted "approval" to "exemplary" sites of reform.

SUMMARY AND CRITICAL ISSUES

In this chapter we have described how we came to terms with the practical and theoretical challenges inherent in collaborative, cross-disciplinary research. We recognize that each study has its own unique methodological features and consequently we have provided no map for future efforts of this sort. Nonetheless, we have identified three critical issues that likely extend beyond the context of our project and that may be of use to other researchers considering a cross-disciplinary team approach to field-based inquiry.

The relationship between researchers' conceptual frameworks and the questions they ask. In retrospect, we were not as deliberate as we should have been in our efforts to identify and deal with the distinct theoretical and conceptual concerns of researchers on our team. As a result, we were inadequately prepared to handle issues of data selectivity and implicit theoretical bias on the part of each researcher. Were we to begin again, we would devote a significant amount of time to systematically seeking out our distinctive disciplinary biases—not retrospectively when the data had been collected and the analysis was well underway, but while our field work was in its formative stages. It is less threatening and more conducive to team-building if the strengths and limitations of individual perspectives are discussed before they are "weighted down" by application to actual data.

The relationship between researchers' conceptual frameworks and how they define "meaningful units" of analysis. Data analysis was facilitated by our fortunate, if not totally deliberate, decision to operate within a range of acceptable interpretative frameworks rather than to impose a single theory of "best fit." This approach mirrored our overall notion of complementary functioning as a research team. This did not entirely resolve differences between those researchers committed to pursuing the complexity of phenomena and those who sought to reduce variables of study, but it did clarify for us that different types of decision making are required for different phases of the research process. As similarly noted by Liggett et al. (1994), consensus seemed a particularly appropriate "rule of thumb" when establishing overall direction in the study. Compromise shifted into play when strong positions were held that impacted individual presentations but did not alter the established aims of the project. Liggett et al. also advised that "team members should be prepared to go forward [in certain matters], not because they agree with one another, but simply in the interest of getting on with the project" (1994, p. 86). In short, it is better (in the sense of achieving a richer and more varied result) to allow for individual latitude than to get bogged down in fine-grained detail.

The negotiation of accounts among researchers with different disciplinary perspectives. According to what criteria do we judge a qualitative study "complete?" How do we negotiate a balance between individual disciplinary voices and an integrated account? Our success in dealing with these questions reflected our ability to resist making pronouncements of "what ought to have been" and, instead, to maintain our commitment to reporting "what was." And while there is an evaluative dimension to all description (Wolcott, 1990b), the "antidote," as Wolcott suggested and we experienced, entails restraint and straightforward acknowledgment of those personal and professional judgments that do creep into one's work. At a basic level, this process required our critical framing or editing of such words as "ought," "should," or "must" in our early drafts. More deeply, it necessitated our participation in a reflective and discipline-spanning dialogue aimed at clarifying how we perceived, individually and as a group, the distinctions among description, analysis, interpretation, and evaluation. The extent to which we were effective in reconciling these perceptions determined in large measure the collective positive impact of our written accounts on practice.

Chapter 3

Institutionalizing Mathematics Education Reform: Vision, Leadership, and the *Standards*

Beverly J. Ferrucci

"Oh, no! Something is wrong," declares a student as she tries to explain her concern to the three other classmates in her group. "Look, we have a rectangle that is one square down and twenty-four across (see Figure 3.1). It's easy to count the squares and see that the area is twenty-four. But getting a perimeter of forty-nine is not a good answer."

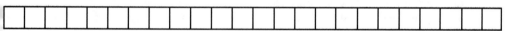

Figure 3.1. Perimeter-area problem

"What do you mean?" asked another student, "It sounds okay to me. I counted the sides of the squares in the picture and it came out to be forty-nine." "Well, look," replied the first student, "We have one square down this side—that's an odd number—and we have another one down the other side—that's another odd number. And one plus one equals two, an even number. And we have twenty-four squares across the top and twenty-four squares across the bottom. They are both even numbers. So, how can our answer be equal to an odd number?" "I don't know," responded the student. "I counted them and it came out to be forty-nine." "But doesn't that bother you?" exclaimed the first student. "Why should it?" responded the second. Shaking her head in bewilderment, the first student replied, "I don't know why. I just know that it bothers me a lot."

This brief vignette is representative of the problem-solving atmosphere that is sustained not only in mathematics but throughout the entire curriculum at Deep Brook Elementary School. This problem-solving emphasis served both to frame and to focus mathematics reform efforts at the building level. My aim in this case study is to explore the context in which these efforts have been embedded. I focus my description and analysis on the dynamics of a school district willing to allocate substantial resources to professional development, a leader (the principal) who encouraged and supported his staff in their attempts to define a mathematical vision to implement instructional innovations, and the role of the NCTM *Standards* as a means of introducing and justifying these changes.

The Deep Brook Elementary School is one of three elementary schools in a rural district of approximately 2000 students. The school houses grades K–5, whereas the second elementary school houses grades 2–5 and the third consists of Grades K–1. There is one high school consisting of Grades 9–12 and one middle school consisting of Grades 6–8.

As discussed in subsequent sections of this chapter, resources and relationships appeared to be consistent with a school-wide vision that placed a high priority on long-term staff development and training in mathematics programs. Other decisions reflected this priority as well. A joint community-school decision resulted in the elimination of the school's hot lunch program, thereby freeing up additional funds to bolster instruction, training, and materials.

In the following section, I describe the mathematical vision that has fostered the problem-solving environment of the school. In subsequent sections I describe the visionary leader who helped make the whole process a reality, the role of professional development, and the validating effect of the *Standards*. In the concluding section I consider the replicability of the Deep Brook School's program, the problems associated with trying to implement and sustain such a program, and suggestions for districts that do not have the financial resources to institute such a program.

THE VISION

A school atmosphere in which each classroom is a problem-solving environment best describes the vision of the administrators and the staff of the Deep Brook School District. They believe this vision can be brought to fruition by forming a vibrant mathematical community that promotes problem-solving skills through exploration, conjectures, connections, data analysis, verification, and mathematical discourse.

As described by Deep Brook's teachers and principal, a child's ability to learn is defined by the cognitive abilities available at different phases of development. Thus children learn best by actively participating, and hands-on experiences are crucial to the development of understanding. These beliefs underlie the teachers' view that students are at the center of all school-based activities and should be viewed as both producers and consumers of knowledge. As one teacher stated,

> They [the students] seem to want to know more. They have more questions about things. I find that children don't want the answers given to them anymore. They are much more active now because they are involved in every lesson. They like the involvement because hands-on is the way to teach because I'm not teaching them. I'm not the center of attention—they are. They're doing; they're asking; and they're getting involved. So I see a happier, more curious, and more problem-solving-oriented student.

Effective mathematics teaching, according to the Deep Brook's administration and staff, occurs when both teachers and students are allowed to take risks and experiment with new techniques and materials. Teachers are encouraged to become risk-takers and to try new ideas and mathematics projects. In turn, they encourage their students to become risk-takers and to feel secure in volunteering responses without fear of ridicule or censure for incorrect answers. As a result, the students at the Deep Brook School have developed enough confidence in their perceptions of mathematics that they feel comfortable hypothesizing, estimating, investigating, and verifying. A first-grade teacher described her students' abilities to solve problems as follows:

> The children are risk-takers. They can verbalize and communicate an answer or strategy. They know that there are strategies, know that there are options, and are willing to try them if they don't get the correct answer at first.

Deep Brook School students are encouraged by their teachers to articulate their mathematical thinking in an oral and written manner. Students at all grade levels keep math journals. Even kindergarten students maintain a class journal in which the teacher records their emerging concepts of mathematics. Students in other grades document explanations of their solutions and strategies. They seem to know how to select appropriate heuristics, collect and analyze data, and verify their results. These abilities reflect a broader vision, in which students are encouraged to engage in mathematical discourse in a problem-solving environment.

Evidence of this discourse can be seen in the following excerpt from a fourth-grade mathematics and science lesson on crayfish. Students were participating in a class discussion and simultaneously recording their observations and results in their journals.

Teacher: Today, each of you is going to receive your own crayfish. I want you to base all your predictions on your own crayfish.... Let's start by estimating the weight of each crayfish in grams. Pick up your crayfish and write your estimate in your journal. After you have made your guesses, write about the ease or difficulty of doing this part of the activity. Then share your ideas with the class.

Students wrote their estimates and comments in their journals. One student summarized the feeling of most of his classmates.

Student: This was hard to do. It would have been easier if we had something to compare to a gram. You know, if I could hold something which weighed a gram in one hand and my crayfish in the other. That would have been a lot easier.

Teacher: Good point! Let's first record the estimates on a class chart. Then I will give each of you a sugar cube which weighs one gram. Use it as a benchmark to give a new estimate of the weight. (Pause) Did your estimates go up, down, or remain the same? Explain what made you change or not change your mind.

The students recorded their new estimates on another class chart, weighed each crayfish, and discussed why they had changed or kept their estimates the same.

Teacher: Okay, let's estimate their length in centimeters now.

A sigh of relief was heard in the classroom as several students commented that length was much easier to estimate than weight. They estimated and recorded the length of the crayfish, used a benchmark of one centimeter to readjust their estimates, and recorded the actual lengths of the crayfish in their journals. At the conclusion of the lesson, the students were given a series of questions to answer in their journals in preparation for the next lesson. They are as follows:

1. Predict the heaviest object you think your crayfish will be able to hold in its claw. Explain your answer.

2. Is the heaviest crayfish, the one that weighs the most, also going to be the strongest? Explain your answer.

3. Make a generalization about Question 2. Then be prepared to defend or reject your generalization, depending on tomorrow's class results.

In striving to meet the goals of the district's mathematical vision, the Deep Brook School staff has organized itself in the following manner. Classroom teachers meet with special education teachers and teacher assistants in order to address the varied problem-solving abilities of the students in their classes. Within the structure of the school day, teachers of the same grade level are provided with a block of common planning time at least three times per week so they can design specific lessons as a group, create student projects, share uses of manipulatives and technology, and collaborate on themes and activities to be used in all their classrooms.

This collaboration has a twofold purpose. It not only enables teachers to share lessons and materials with each other but serves in a supportive role as well. In 1992 they started peer coaching, a method whereby colleagues are invited into classrooms to observe lessons and offer suggestions and encouragement for improvement.

An outcome of these collaborative efforts is that teachers have become more confident in their understanding of mathematical topics. A teacher described in the following way her enthusiasm for teaching mathematics and her mathematical understanding after attending workshops:

> When I first started taking the workshops, I started seeing math in a different way. I started understanding things that I had only memorized before. If at my age I'm just realizing these things, wouldn't it be nice if children could see that from the beginning—if they could realize that there are lots of different ways to solve a problem and how to verbalize the way to solve a problem?

This confidence inspired the teachers to redesign the school's mathematics curriculum. A consequence of this redesign has been to a large extent the abandonment of the use of textbooks in determining the content of the mathematics curriculum. A few teachers in the upper grades do continue to use textbooks, but the teachers in the lower grades have opted to discontinue the widespread use of textbooks. This approach could be viewed as consistent with the view offered in the *Professional Standards for Teaching Mathematics* (NCTM, 1991):

> textbooks can be useful resources for teachers, but teachers must also be free to adapt or depart from texts if students' ideas and conjectures are to help shape teachers' navigation of the content. The tasks in which students engage must encourage them to reason about mathematical ideas, to make connections, and to formulate, grapple with, and solve problems. Students also need skills. Good tasks nest skill development in the context of problem solving. (p. 32)

Although the Deep Brook School teachers agreed that this type of program requires more planning time because of the lack of textbooks, they remain no less passionate about their mathematics program. They realize that they must supply the students with problems, projects, and related activities. However, they feel that this lack of dependency on textbooks forces them to collaborate more with each other and be more creative in their teaching.

Another important component of the vision is an emphasis on the integration of mathematics with other subject areas (e.g., social studies, science, language arts) in the context of real-world problems. The superintendent noted that the district's teachers were constantly seeking to create a variety of challenges for students that place them in a situation where they have to solve complex, authentic tasks.

One such task occurred during a science unit on fossils and dinosaurs. Second graders were asked to devise a method of illustrating the length of a dinosaur. The students used reference books to determine the approximate length of the dinosaur, recorded measurements on the hallway floor, placed strips of tape the desired distance apart, and walked on the tape to get a feeling for the actual length of the dinosaur.

The Deep Brook School teachers believe that these types of lessons are much more meaningful to the students than pencil-and-paper activities. A parent commented on how well her daughter was able to relate to the dinosaur lesson:

> My second grader came home yesterday and she was all excited. She had brought in a banana for snack and the teacher had measured the banana and associated that with the size of a dinosaur tooth. She could relate to that.

One of the most striking aspects the researchers perceived of the vision of reform implemented by the Deep Brook School was the change in how students viewed the role of the teacher. Students no longer saw the teacher's role as one of a lecturer or dispenser of information but rather as one of questioner and facilitator of learning. It is not uncommon for the Deep Brook School teachers to begin their classes with a mathematical question or problem that the students are to investigate for the class period or for a more extensive period of time. A host of resources—manipulatives, textbooks, reference books, markers, graph paper, and technology—are readily available to the students to help them in their problem-solving explorations. Perhaps more importantly, the students recognize that they are valuable resources for each other and often discuss their ideas and strategies amongst themselves in order to collectively draw conclusions about their thinking processes. An example of students acting as both risk-takers and resources to each other is illustrated in the following classroom vignette.

> "Potatoes, potatoes, everywhere! What are we going to do with them?" asks a first grader as she enters her classroom and discovers potatoes lying on the students' tables and the teacher's desk. "We are going to become potato farmers for a week," replies the teacher.
>
> For the next few days, the students estimated the circumference of potatoes, discussed various methods of weighing them, conducted a survey of their favorite ways to prepare potatoes, graphed data from their survey, and used money to purchase materials to design potato creatures.

During one lesson, as the students sat in a circle on a carpeted area with five potatoes neatly placed in a pile in the center, the teacher said to them,"We must weigh our potatoes today with our classroom materials. We want to be able to have a fair measurement so we can see which group's potato is the heaviest. Can you think of some ways to weigh our potatoes?"

After a short debate, the first graders agreed to use a plastic balance scale with the potato on one side and small plastic teddy bears on the other. Based on their obvious familiarity with the technique, it was a procedure they had previously used in class. As the students were being arranged into groups of three or four, they each held their group's potato and gave an estimate of the number of teddy bears needed to balance the potato.

Although each group was using the same materials, they approached the problem in different ways. Three of the groups put the potato on one side of the scale and placed teddy bears one by one on the other side until it balanced. Another group placed their potato on one side of the scale and emptied the entire box of teddy bears onto the other side. Then they began to remove the teddy bears one by one until the scale balanced. The final group placed the teddy bears in groups of five. They began by putting 25 teddy bears on the scale and then added or subtracted groups of five to save time. When they came close to balancing the scale, they began to add or subtract the teddy bears one by one. Each group then counted its teddy bears and found a weight for each potato.

The students returned to the circle to discuss their findings and to record and graph them on a poster board. As the weights of the potatoes were listed in ascending order from lightest to heaviest, a dilemma arose. "Something is strange here!" exclaimed one student. "What do you mean?" asked the teacher. "That can't be right," interjected another student before the first student could reply, "Look at how small that potato is, and look how big *that* potato is. The sizes are different, but the graph says they both weigh 24 teddy bears."

This observation of different-sized potatoes having equal weight sparked a series of possible explanations by the students. "There's just so much stuff inside that one, you can't see it. It's got to be really, really tight together in there," responded one student. Another student added, "Well, you know a softball is bigger than a hardball, but a hardball weighs more."

The teacher offered this insight: "Well, how can we find out if they really are the same?" She appeared pleased with the class's response of putting one potato on each side of the balance scale. However, the two potatoes did not balance the scale. One potato was clearly heavier than the other.

The heavier potato was reweighed on the original scale and the number of teddy bears was found to be correct. However, when it was weighed on one of the other scales the number of teddy bears increased considerably. After checking the scales and noting that they were functioning properly, the teacher appeared momentarily to be at a loss for an explanation. She quickly redirected the question back to the students, "How can this be happening? Can we come up with a reason?"

The students returned to their groups to try to solve this mystery. For several minutes the groups discussed the problem and hypothesized about the discrepancy. One by one the groups asked questions about the potatoes and offered their hypotheses. The other students listened to their explanations and gave reasons why they thought their theories were incorrect. Finally one student asked a vital question, "Were the teddy bears all bought at the same store?" The teacher explained,

We order our materials from companies and they send them to us. So most of the time we don't go into stores to buy them. But that's a good question. Let's look at the teddy bears' storage boxes and see where they came from.

On inspection it was found that the teddy bears were indeed from different companies, with one brand weighing slightly more than the other. This made it necessary to weigh the potatoes again using the same brand of teddy bears for all measurements. With the new calculations in hand, the students regraphed the data and drew new conclusions.

The use of such innovative approaches and materials in mathematics lessons, according to the Deep Brook teachers, helped their students to recognize the importance of mathematics, demonstrated the students' willingness to spend a great deal of time working on mathematical tasks, and developed the students' abilities to become risk-takers, problem solvers, and resources for each other.

In order to develop a strategic plan that would result in teachers using more innovative approaches to teaching mathematics, the district administrators felt that an ongoing professional development program should be established. They also believed that they themselves must become knowledgeable of the current trends in mathematics education as well as the obstacles that might inhibit more extensive changes from taking place in mathematics instruction in the district's schools. This theme is the focus of the next section.

LEADERSHIP

I think, number one, you need an administrator who understands—not just parrots phrases but really understands—what mathematics is and how children learn mathematics. Then you need a district that's committed to feeling that the training and support of teachers are paramount.

—Third-grade teacher,
Deep Brook School

These feelings, expressed by the majority of the teachers interviewed, describe the motivational role of the principal and the supportive role of the school district in effecting reform. These two factors had a significant impact on the mathematics reform efforts of the Deep Brook School District.

The principal of the Deep Brook School was the major catalyst for mathematics education reform not only in the Deep Brook School but throughout the district. Because of his efforts, professional development activities were a priority for the eight years prior to the site visits. He routinely volunteered to model appropriate mathematics instruction at all grade levels within the school and lead workshops on mathematics programs and was the vehicle through which the teachers became knowledgeable about the *Standards*. In addition, he spoke at local, state, and national mathematics education conferences and served as a resource for mathematics reform initiatives for schools throughout the region. The prominent role played by the principal was reinforced by the superintendent:

> Clearly the most significant factor has been the building principals who have become knowledgeable in the areas of mathematics and mathematics instruction; who have become knowledgeable in staff development and adult learning processes; who have implemented a staff development program; and who have communicated a vision of mathematics on a long-term basis to the staff and the constituents in the schools.

The principal expressed his belief that teachers should assume a leadership role in any change that occurs in the programs within the school. He acted as an agent of change, but also realized that if any change in instruction is to be effective it must be initiated by the constituents themselves. Thus, he advocated the transfer of ownership of the program that involved changing from a prescribed curriculum adopted by the Board of Education to one influenced by the teachers and the students.

The principal was complimented by the teachers for his dissemination efforts, which included providing research and other professional writings in his interactions with them individually and collectively. He often placed copies of articles and documents in their mailboxes on a weekly basis. These readings appeared to be helpful, because the teachers frequently referred to them as they collaborated on the redesign of the curriculum. Announcements of upcoming workshop opportunities, conferences, and new ideas and themes in mathematics education were also announced at faculty meetings. The pivotal role of the Deep Brook School principal in mathematics reform efforts cannot be overemphasized.

All 21 Deep Brook classroom teachers and most special-area teachers (teachers of music, art, physical education, etc.) willingly participated in ongoing staff development workshops. These in-service opportunities were made available by their school district, which had made a commitment to allocate significant funds to providing a formidable professional development program for the staff. In addition, the district formed a working relationship with mathematics educators from the Mathematics Learning Center at Portland State University in Oregon. These mathematics educators frequently conducted workshops for the teachers and administrators in the Deep Brook School District and allowed them to participate as instructor-trainees during their presentations. This opportunity allowed for the Deep Brook teachers and administrators to serve as future in-district resource instructors for their colleagues, both in and outside the district. Furthermore, school administrators encouraged teachers to pursue mathematics training at the college level.

At first, not all teachers were willing to participate in the in-service workshops. However, after participating in the mathematics staff development activities, teachers began to discuss concerns and share materials and ideas. The enthusiasm began to spread, and other teachers who had not attended the workshops were curious to learn about them. They wanted to know what to do differently in their classrooms that would result in an improved attitude toward mathematics and a more in-depth approach to problem solving. The principal described the teachers' initial reactions to the workshops as follows.

> I had a very enthusiastic response from the staff. Once the initial people got excited about it, they began to spread the word among other staff members. It took root and things began to grow beyond that. Somebody else learned about it, saw what was going on, liked it, and began to incorporate it.

Since then, interest in the workshops and other in-service opportunities prolif-erated. At the time of our study there had been up to 280 teachers from the Deep Brook School District and surrounding districts enrolled in program sessions at one time.

Reflecting on the process of change and the decision to invest the district's money in providing mathematics instruction and workshops, the principal said,

> I think that anyone who expects to bring about change has to be very serious about it, very committed to it, has to have thought it through very carefully, has to have planned many opportunities for people along the way, has to anticipate all kinds of setbacks, and has to alternate strategies thought out before the fact—and has to be able to do it in a way that indicates to people who control money that it's not as expensive as it may appear, that the results it is going to bear are going to be better than the results you are getting with whatever you are doing now.

THE *STANDARDS*

The *Curriculum and Evaluation Standards for School Mathematics* and the *Professional Standards for Teaching Mathematics* have acted as resource guides and validating tools, in varying degrees, for the Deep Brook School teachers and have had an impact on the way they conduct their mathematics teaching and organize their classrooms.

The *Curriculum and Evaluation Standards for School Mathematics* served as the justification for the Deep Brook School teachers to develop a curriculum in which problem solving, connections, communication, and mathematical reason-ing are central themes. The teachers adopted this philosophy and strove to create a classroom environment that encompasses all these factors. The nature of this environment was shaped by the types of mathematical tasks presented and the ensuing discourse in which the students engaged. Teachers constantly tried to perfect their skills by developing and integrating worthwhile mathematical tasks into their instruction. This view of learning is also in accord with the *Professional Standards for Teaching Mathematics*:

> Students' opportunities to learn mathematics are a function of the setting and the kinds of tasks and discourse in which they participate. What students learn—about particular concepts and procedures as well as thinking mathematically—depends on the ways in which they engage in mathematical activity in their classrooms. Their dispositions toward mathematics are also shaped by such experiences. Consequently, the goal of developing students' mathematical power requires careful attention to

pedagogy as well as to curriculum. (p. 21)

In selecting, adapting, and designing mathematical tasks for their classrooms, the Deep Brook School teachers based their instruction on five major shifts in the composition of mathematics classrooms advocated in the *Professional Standards for Teaching Mathematics*. These shifts are necessary if teachers are to move from classrooms that concentrate on the mastery of computation to ones in which students construct their own mathematical understandings. These shifts are (1) toward classrooms as mathematical communities—away from classrooms as simply a collection of individuals; (2) toward logic and mathematical evidence as verification—away from the teacher as sole authority for right answers; (3) toward mathematical reasoning—away from merely memorizing procedures; (4) toward conjecturing, inventing, and problem solving—away from an emphasis on mechanistic answer-finding; and (5) toward connecting mathematics, its ideas, and its applications—away from treating mathematics as a body of isolated concepts and procedures. (p. 3)

The majority of the Deep Brook teachers attended workshops, in-service meetings, and mathematics conferences that dealt with ways in which to interpret the main themes from the *Standards* for use in their classrooms. In interviews with outside documenters, the teachers stated that they felt their mathematical vision was verified by key features of the *Standards*. In fact, most of them felt they had possessed a teaching philosophy similar to that proposed in the *Standards* prior to its publication. To them, the four strands found throughout the *Standards*—problem solving, communication, connections, and mathematical reasoning—were already part of their repertoire.

The Deep Brook teachers had been participating in professional development activities in mathematics and modeling a problem-solving atmosphere in their classrooms for the 8 years prior to the 1993 site visits—well before the release of the *Standards*. For them, the vision had passed the interpretation stage and was now in the practice stage. At the time of this writing they were in a stage of continued reflection concerning the *Standards* looking for opportunities to extend their efforts at reform.

DISCUSSION

The Deep Brook School had a core of teachers who were able to devise and sustain a strong mathematics curriculum, with the expectation that the excitement and curiosity generated by the mathematics program would be transmitted positively to students, parents, and others in the school community. To help solidify the infrastructure of the school, the principal aggressively sought to increase parental involvement in all of the school's activities. The administration sustained the belief that parents are a vital part of the process of change. Parents were invited to participate in the educational activities of their children and provided with suggestions and actual activities that could be done at home as a family unit. These activities not only reinforced what the children had learned in school, but also allowed parents to become more aware of the curriculum and mathematical content their children were learning.

Deep Brook School's principal needed the endorsement of the participating teachers to meet his vision of creating a problem-solving environment throughout the entire school. He began to talk to them about the themes of the *Standards* and to model appropriate mathematics classes for them. In turn, the teachers required a well-planned professional development program to assist them in instructional and methodological areas such as learning more about the *Standards*, updating their mathematical knowledge, and developing skills for using manipulatives and technology in a fully integrated environment.

However, this approach was not without its problems and challenges. The teachers at the Deep Brook School have not arrived at their present position of reform without prior difficulties. As with any changes or major reorganization, change was difficult to achieve at first, if at all. As one teacher stated,

> It's hard to change! It's much easier to stay with what you know; you feel comfortable with that. So I think [you should] take it easy on yourself and do one thing at a time and then feel like you've accomplished that and then move something else in all the time. Get that area so that you feel good about it, next year add another area, then another area, then another area.

This sentiment, expressed by a number of Deep Brook School teachers, is one factor that these teachers believed facilitated the transition from a computation-oriented curriculum to a problem-solving program. The Deep Brook School teachers felt firmly that change should be made gradually and in small increments. The principal echoed the same notion:

> Don't expect to implement the entire program in one year. It is just not possible to do. It takes several years for [teachers] to get everything internalized and integrated and so part of the message that we communicate to them and try to reinforce is that you take what you understand and what you can manage effectively and work on that, even if it's only one thing.

During interviews with the outside researchers, the teachers reinforced this concept and expressed the importance of relating to others the amount of time it requires to put such a program in place. As one first-grade teacher noted, "They need to be reassured that it does not happen overnight or even in one year. It must be sustained over a period of time to be effective."

In general, Deep Brook's success serves well to remind us that teachers must do more than share and coordinate content—they must clarify their own mathematical and pedagogical visions. They must meet the challenge of providing students with the opportunity to experience important aspects of mathematical theory and knowledge in an applied setting. They must also explore how mathematics as a discipline is benefited or compromised by integration.

Teachers whose methods of instruction have been fashioned by years of lecturing and diagnostic teaching may find it too difficult to abandon a familiar routine in favor of a more constructivist approach. For them, playing the role of facilitator rather than lecturer places them in an uncomfortable situation. The

Deep Brook School teachers indicated that the process of change should not be undertaken alone. They attributed part of the success of their program to colleagues who encouraged and supported their attempts to change.

Professional development, as an important component of a school district's plan to improve instruction, is not always given adequate attention by administrators in times of changing mathematics education. Many districts, facing diminishing financial resources, are prone to reduce expenditures for in-service opportunities as a common cost-saving measure. As an alternative, districts might consider implementing a program that will train a small cadre of teachers, or even a single teacher, who could become resource specialists in mathematics and begin training other teachers in the district. Having teachers attend local mathematics conferences and requiring them to make presentations upon their return might also be beneficial. Another option is to form a partnership with a local university or college that can involve teachers in its own preservice and in-service teacher preparation activities or involve them in the implementation of a funded professional development program.

Two key assumptions from the *Professional Standards for Teaching Mathematics* succinctly summarize the reform efforts at the Deep Brook School: (1) teachers are key figures in changing the ways in which mathematics is taught and learned in schools, and (2) such changes require that teachers have long-term support and adequate resources (p. 2). The teaching envisioned in both of the *Standards* documents is significantly different from the experiences current teachers witnessed themselves as students in mathematics classes. Understandably, it will take teachers time to change their role and the nature of the classroom environment. Ongoing professional development coupled with encouragement and support from administrators, colleagues, and parents were crucial elements for success at the Deep Brook School.

CONCLUDING REMARKS

Historically, mathematics learning has often taken place within an atmosphere of rigidity and student fear, in classrooms void of participation, discourse, and active investigations. This was definitely not the case at the Deep Brook School. The teachers moved beyond how they were taught and instead created a classroom atmosphere in which students were actively participating in a variety of challenging and worthwhile mathematical explorations. They brought to life their vision of the mathematics classroom in which they wish they had been taught.

By initially providing teachers with a multitude of professional development opportunities, the district was able to lay a solid foundation in mathematics within the infrastructure of the school. The workshops and other in-service activities had a powerful impact on both the teachers' teaching and their self-esteem. At the time of our study, teachers appeared quite confident and comfortable experimenting with complex mathematical tasks in their classrooms. This resulted in positive attitudes that reverberated throughout the school and community.

Our findings suggest that the reform enacted at Deep Brook School and supported by community resources were indicative of a collective vision of the teachers and were changes of substance, not merely style. The strength of this claim is being tested at the time of this writing. During the period of documentation, the Deep Brook School District was facing a $4.8 million reduction in revenues because of cutbacks in the high-tech industry in the area.

We realize that many factors have influenced mathematics reform in the Deep Brook School District, not the least of which was the Deep Brook School principal. He emerged from our data as a significant force in the reform efforts. His aspirations for an investment in teacher change at the elementary level eventually became the catalyst for change at the district level. As a result, the Deep Brook School teachers have become knowledgeable resources not only within their own district level but in neighboring districts, as well.

This research provides teacher educators, administrators, curriculum supervisors and classroom teachers with information on how one school integrated a problem-solving approach into its overall teaching vision. It identifies factors that may have had an impact on the teachers' attitudes toward mathematics and mathematics instruction. In addition, the study holds promise for documenting the evolution of change when a vision is articulated, strong leadership is established, and the *Standards* are used to guide and complement the beliefs and philosophy of a school district.

Chapter 4

Walking Together on Separate Paths: Mathematics Reform at Desert View

Thomas Schram and *Loren Johnson*

"HEAR YE! HEAR YE! HEAR YE!" proclaimed the teacher imperiously. "From this point on, I am Prince Navarjo!" The 25 students in the honors geometry class, a balanced mix of male and female, White and Hispanic, sat with mouths agape at the unexpected antics of their instructor. Their initial shock gradually shifted to laughter and a shaking of heads as "Prince Navarjo" distributed ribbon-enclosed "decrees" and continued, "See to it that you complete your tasks in a timely manner and with great pride, or I will see to it that you rot in the dungeon!"

Provided with a diagram of a castle's exterior and the information that one gallon of paint can cover 550 square feet, each group of three or four students confronted a multilayered task: to find the total lateral surface area of the castle to be painted in celebration of the king's [the school principal's] arrival; to advise Prince Navarjo how many cans of paint to purchase; and, by computing the total surface area of the roofs on the towers and main building, to determine whether the Prince has enough money to repair them. The teacher informed the students that some class time would be provided over the next 2 to 3 weeks to complete this task, termed a *project* by the university mathematicians who originally devised this curricular approach and by the local high school teachers who adopted and adapted it.

This brief vignette is representative of numerous classroom encounters that, combined with documentation of the broader school and community context, have contributed to this study. Taken together, these encounters and the perceptions of those who participated in them highlight the complex interplay between content and process, leadership and ownership, risktaking and routine. The broad aim of this paper is to invite reflection on these issues as they emerged and shaped our study's central theme: how teachers, administrators, students, and university faculty create the climate in which an instructional innovation is conceived and then transformed into a vehicle for ongoing reform in mathematics education.

At the time of our fieldwork in 1993, Desert View High School served approximately 2000 students, roughly one third of whom were classified as low-income—in many cases living at or below the poverty level. An even mix of White and Hispanic, the student population reflected the school's proximity to the Mexican border and necessitated a comprehensive English as a Second

Language (ESL) program. Desert View's position as one of three high schools in a large but financially strapped district limited the availability of instructional resources and heightened reliance on external grants. During the 1993–94 academic year, for example, Desert View's 13-member mathematics department shared a $1000 budget for materials.

Questions of how to, on the one hand, tailor instruction for students who had never experienced success in mathematics and how, on the other, to prepare academically motivated students for college were primary concerns of Desert View's mathematics teachers. Teachers and administrators alike cited student absenteeism as the major obstacle to effective teaching and learning, noting in conjunction with this problem that about 16% of the ninth graders dropped out each year. The school-wide dropout rate in the years 1988–1993 averaged nearly 9%. Even so, analysis of enrollment patterns between 1990 and 1994 revealed an increase in the number of students in higher level mathematics courses (Algebra II and above) and a dramatic decrease in the number of students enrolled in remedial courses. The number of remedial class sections decreased from 33 in 1990–91 to 12 in 1993–94, whereas the number of class sections offering Algebra II or higher classes increased from 30 to 33. Over the same period, the average rate of student failure in mathematics—though largely a reflection of student attendance patterns—decreased slightly from 15.5% to 14.9%. Desert View teachers and administrators perceived the net result of such trends as positive. In the words of one teacher, "We're teaching a lot more math successfully to a lot more kids, getting more of them involved in the math mainstream than ever before." In this chapter, we focus on the substance of this change and the conditions under which it took place.

In the following two sections we describe the formative influence of a partnership between Desert View's teachers and members of the mathematics faculty from the local university. We direct particular attention to the creation and implementation of the project approach in university classes and its deliberate reconfiguration in the high school setting. Subsequent sections provide description of "project-driven" reform in the high school mathematics program, with emphasis on patterns of teacher-teacher and teacher-student interaction; attempts to address the instructional needs of at-risk students; and reconsideration of what is meant by "doing mathematics." Our analysis in the final section highlights the promising and problematic nature of shared leadership amidst individual variation in the reform process.

Our account emphasizes the importance of contexts constructed by Desert View's teachers and students as they confronted changes in pedagogy and self-perception and then translated these changes into new ways of applying and communicating mathematical understandings. We highlight selected issues and events for their value in suggesting processes that can be fostered in other reform-minded schools—though not in the exact configuration found at Desert View.

FIRST STEPS

One thing I've learned from this experience [with project-driven reform] is that you can take a fairly traditional high school in a very traditional setting and you can break some ground, you can do some nontraditional things that have not been imposed by edict or force. No one said that this had to happen.

—Assistant Principal,
Desert View High School

In 1988, a group of mathematicians at the local university, supported by a National Science Foundation grant, began to experiment with a calculus curriculum that involved the use of projects—extensive multilayered problems that students had to solve and then explain in technical reports. The projects emphasized group work, including some cooperative learning, and the ability to express one's thinking in written form. For the "pure mathematicians" at the university, the project approach represented not simply a new curriculum but a new mindset for mathematics instruction. One of the program's main supporters described this as a shift toward greater student ownership of the learning process:

We gave [the students] a reason to become involved with mathematics. One of the biggest things we did was change the time frame in which students solve mathematics problems. It used to be they could solve 5 to 20 of them in an evening. Now it's one problem that takes a week or two—so you view the whole problem differently. You start out with something you absolutely can't do, and a little while later you end up with something that you can do.

The same year, a small group of secondary school teachers in the local school district were completing their fourth year of monthly meetings as a mathematics curriculum committee. Originally formed in 1984 for the purpose of curriculum revision and textbook adoption, the committee members—a large proportion of whom were from Desert View High School—chose to maintain the group as an ongoing forum for the exchange of ideas among mathematics teachers in the district. The committee continued to have a major voice in the district's program and staff development decisions at the time of our fieldwork.

Even with these structures in place, early contacts between the university and the Desert View schools were more serendipitous than systematic. In the summer of 1990, complementary concerns brought together university mathematicians with teachers from the mathematics committee, in particular, several from Desert View High School. The university personnel were motivated by the desire to improve the preparation of students entering their program, particularly in terms of their ability to solve problems and write technical reports. More to the point, several of the mathematics professors sought an effective means of socializing students into the project approach. "Quite frankly," noted one professor, "we wanted to avoid the battle with students over whether they should write in mathematics class. We thought if they did this in high school, they would see it as a legitimate way to do work when they got to the university." One of Desert View's teachers suggested, more bluntly, "The university people wanted better projects without having to teach [the students] how to do them—but it clearly met our desire for instructional innovation as well."

For their part—at this point quite separate from the issue of projects—the high school teachers sought enhanced communication with the university:

> We wanted them to be interested in us—more than how many students are we going to send them next year. We wanted them to provide us some guidance and expertise. We knew that we were training kids to the best of our ability in mathematics, but that wasn't necessarily what people at the university wanted. We had a communication breakdown between what we envisioned students needed according to SAT tests and state competency exams, what we did in high school, and what people on the cutting edge of mathematics needed from us now. We had a big reality gap that we needed to close. It seemed natural to work with the local university.

The professors at the university introduced the idea of using the project approach to teach high school mathematics. Many teachers expressed early concerns about this proposal, citing fear of writing, of grading such a nebulous kind of homework, fear of hostility and rejection on the part of students, and fear that they themselves might fail or have to abandon the program during the year. Others promoted the benefits of engaging students in practical, in-depth problems. Reflecting this cautious optimism, one teacher commented,

> I thought the whole concept was wonderful. I knew that it would be difficult to convince some of the teachers who were set in their ways to try it. But as a new teacher, I was always looking for something else to make mathematics more applicable, and I knew that the projects did that. They took something mathematical and converted it to something real, rather than just a lesson in the book. I knew it would do that for us. My only concern was whether we could write projects at the [high school] kids' level that were real-world applications and not just fluffy, "busy work" things to do.

With additional funding from the National Science Foundation, local university mathematicians designed a program to help teachers design projects that could be incorporated into the high school mathematics curriculum, with an initial focus on Algebra II, geometry, and trigonometry. Over the next two years, an increasing number of Desert View teachers chose to become involved in the writing, editing, and implemention of projects, extending the strategy, or variations of it, into calculus, Algebra I, and remedial classes. By the fall of 1993, every member of the department used projects in his or her classes to varying degrees and approximately 100 projects were on file in the high school. The projects ranged in scope from developing construction plans for an arch that spans a large river to determining the best day to play a football game by using biorhythm functions.

For a school district with limited funding for in-service programs and consultants, and in which only 2 days per year were officially designated for curriculum development, the collaborative NSF program was a significant boon. "I have to say that [the university's] ability to come up with some funding for an incentive to get the ball rolling on the program and get the teachers to make a commitment for outside time was really instrumental," noted one teacher, echoing the feelings of nearly all the teachers we interviewed. University mathematics personnel conducted summer workshops in which teachers developed and discussed projects. Teachers participating in the initial workshops were given projects to solve that had been used by the university's mathematics department in calculus courses. This was done in part because no projects had been written to date for the high school level, but more importantly, to place teachers in situations in which they would experience mathematical learning in a manner similar to their students. Group work during the workshops was structured to highlight the potential problems and benefits of a cooperative learning framework. Teachers graded each other's work as a means of focusing on their own mathematical vocabulary and as a model for future peer editing among students. (Three years into the program, the question of how to grade projects objectively remained a concern of most Desert View teachers.)

Also instituted were four meetings per year in which university mathematicians and high school teachers could share concerns, strategies, and long-term plans. University mathematicians provided feedback on the mathematical content and quality of teacher-designed projects, although, as noted in the following section, Desert View's teachers quickly assumed ownership of the design and modification of materials to meet the needs of their students in the high school. Those teachers who participated in these activities also received stipends through the grant, although teachers with whom we talked seemed as appreciative of the psychologically supportive role played by university personnel involved in the project as they did of the stipend. As one Desert View teacher remarked,

> Certainly the feedback and interest we have received from the principal investigators [mathematics professors] on the project program has made a big difference. Because we have somebody outside the system who has been interested in what's going on, who has been in communication with us, who has asked provocative questions, we have more pride in what we do. We think of ourselves more as professional educators and not as just holding down the fort day to day .... The principal investigators have always afforded us respect for what we do at our level.

These perceptions match those expressed by the university professors who served as principal investigators in the program. When asked about the respective roles envisioned for the university and high school personnel, one mathematics professor responded,

> We needed the high school people to help us with the logistical part of doing with high school students what we had been doing with college students .... We offered the pedagogy and the mathematical strength and they offered teaching experience and understanding of the society involved. The original proposal was very egalitarian—we have expertise in some stuff and you have expertise in other stuff, and if we put them together maybe we can get a good system going.

PICKING UP THE PACE

> I feel that they [the university people] don't tell us what we have to do; they just offer ideas to us and we adapt things the way we want—and that's okay.
>
> *—Mathematics Teacher,*
> *Desert View High School*

Feelings like those expressed above were expressed by a number of Desert View teachers, contributing to a contrasting set of perspectives on the leadership and direction of project-related reform efforts. As outside documenters, we perceived significant teacher ownership of the program—a perception shared by university mathematicians, who tended to refer to the high school teachers as "the experts in implementation." In contrast, Desert View teachers conveyed the perception that university personnel continued to play the leading role in program design and development. The same teachers, however, commented on the need for the mathematics professors to understand and appreciate more fully the daily reality of public school life—"the nuts and bolts of teaching," noted one district administrator, "everything from how classes are organized and how kids behave to what the textbooks are like."

A number of teachers expressed appreciation for the fact that the mathematics professors devoted significant time to observing and interacting with the high school classes. The university mathematicians' periodic presence in the classrooms is in itself important—even as we consider that their role rarely extended beyond that of an unobtrusive demonstrator of support for each teacher's efforts.

Despite the varied perspectives of high school and university personnel, individuals in both settings noted that once in the hands of Desert View's teachers, the project approach assumed a pace and pattern of implementation different from that which existed at the university. The university's definition of a project as a challenging "multistep, multiday, nontraditional mathematical assignment," for example, assumed a more comprehensive interpretation in the high school. As stated by Desert View's mathematics department cochair,

> The process of incorporating projects into mathematics classes does not just mean giving students longer or more difficult problems. It is a change in the way students do mathematics and science. [Because] students work in groups to analyze a problem, brainstorm answers, break the problem into manageable pieces, and arrive at a solution to which all can agree, the process mimics many job situations. [Because] projects help students teach themselves and each other, the material learned becomes a part of the student, and this gives the student a stake in [his or her] own education. The fact that projects extend and apply the students' knowledge and require them to write clearly and explain their solution means that students remember what they learn and see new connections. The real-life applications of projects awaken many at-risk students to the fact that their education has some bearing on life outside the classroom.

The university personnel we interviewed readily discussed the fact that they were mathematicians, not mathematics educators, and admitted to lacking the insights necessary to work with students in the high school setting. One of the teachers who initiated contact with the university made this frank assessment:

> They told us what they were doing at the university and I didn't like it—and I still don't like the way they do it exactly for my classes because I have younger kids and I have to do it differently .... Out there [at the university], for example, they don't do a story line, which is something most of us do here to make the projects more interesting. I try very hard to write projects that the kids can relate to.

Perhaps most telling were the contrasts we uncovered through our discussions with teachers who had been students in the university's mathematics program and then had worked as student teachers in Desert View High School. Although enthusiastic about the use of projects in both settings, these former student teachers provided insight into the high school teachers' thoughtful re-creation of the project approach to meet the particular needs of their students and their setting. Consider, for example the use of a story line to structure and enhance a project write-up, as noted in the quotation above. The use of a story line required students to place their mathematical problems in the context of a real-life situation and extend their write-ups to include a descriptive, engaging account of that situation, typically in the form of a story. What seemed to Desert View teachers a natural and pedagogically sound means of engaging students at all interest and ability levels in the analysis and writing process was met with some skepticism by the university mathematicians who originally devised the project approach. "I thought it was distracting them [the students] from the mathematics," commented one professor. He explained his reasoning and then described how he came to reconsider the strategy:

> I guess it was [the teachers'] idea that if there was a person in the group who wasn't strong in mathematics, his or her job could be to write the story line for the problem. I said, "Shouldn't the group members take on responsibility for everyone understanding the mathematics? That should be primary." But from talking to the teachers and seeing the student products, I was convinced that some students who wrote story lines learned more mathematics than they would have had they been spoon-fed the material by their peers or the teacher …. I now think it engages people more than it distracts them.
>
> In a recent calculus class I taught, I didn't feel the need to engage the students with a story line—but I said that, for extra credit at the end, they could pose a fictional situation in which this subject might be of use. Some kid wrote about curves in space as shown by the motion of a baseball and the different pitches. A knuckleball has this kind of trajectory and a fastball has this kind of trajectory. It was fantastic! He thought about the angular momentum, and although it wasn't completely rigorous, he now knew some mathematical language that he could apply to a real situation.

More deliberate and extensive patterns of communication and feedback signaled another change introduced by Desert View's teachers. For students who had experienced mathematics instruction in both the university and high school settings, key issues arising from their use of projects included feeling comfortable talking about mathematics, being encouraged to ask questions, and receiving meaningful feedback from both teachers and classmates. Although it was apparent to these individuals that communication and feedback were enhanced at Desert View simply because the high school classes met daily instead of only two or three times a week, as at the university, they attributed much of the success of the new approach to the shared expectation of ongoing feedback and discourse among teachers and students. As one student teacher noted,

> The Desert View High School approach with projects is much more process oriented than the university's [approach] … not only do you have feedback from your teacher, but you also have classmates who are giving feedback. At the university, we

just received a grade at the end of our project. Here at Desert View, students receive extensive feedback on their mathematical problem solving and their writing through-out the project. The university project courses, in comparison, were pretty bare and basic, just one project after another without any real feedback ... it's almost a silent teaching experience. Very little is said; it just happens ... We had one chance to get it right, to make it pretty, and that was it.

After 3 years of program implementation, the stage appears to have been set for a more reciprocal exchange of ideas and instructional innovations between the high school teachers and the university mathematicians. "It's creating a com-munication phenomenon among these technology types who don't know how to talk or how to encourage questions from students," noted one former university student. "And now I see that the students who are in the professors' offices ask-ing questions are the ones who were in the project classes [in high school]." She described the influence of projects on her own university experience.

One of the things I was looking for from my mathematics professors was for them to develop this idea of key questions, of teaching students how to ask the right ques-tions. What is a good question and how do I follow the logic of this idea? Where is this theorem coming from and how does it really fit in? I think that without the pro-jects I would not have understood that as soon as I did, because in writing math, you have to explain math. And if you're explaining math, that means you are also read-ing math—I mean really *reading* math.

More deliberate efforts to expand project-related reform to the middle schools have met with preliminary success. Although the NSF grant funding collabora-tion between the university and Desert View High School ended in the spring of 1993, teachers and administrators described promising steps toward continued district-wide curricular development in the spirit of the original grant. The cochairs of Desert View's mathematics department obtained Eisenhower funds to finance a 1993 summer workshop to help middle school teachers implement the project approach in their classrooms. The funds were also earmarked for future staff development sessions. In addition, a small number of upper-grade elementary school teachers expressed interest about ways in which they might adapt projects for use in their classrooms. Entering the 1993–94 academic year, the district appeared ready to accommodate multiple levels of ongoing project-related reform.

NEW PATHS, NEW VISTAS

Although projects represented only a small proportion of the curriculum (less than half of the average teacher's classroom time per year), they remained the central focus of mathematics reform efforts at Desert View. Projects, in effect, became the "prime movers" of change, prompting increases in mathematics-based writing exercises on a daily or weekly basis; cooperative learning as an instructional strategy; hands-on laboratory activities, particularly in remedial level classes; and peer editing of student work. Especially noteworthy conse-quences of Desert View's involvement with projects were the following out-comes—some incidental, most deliberate.

Peer Coaching and Increased Collaboration

Work on the projects served both to reinforce and to focus communication among the Desert View teachers. As a result of the mediational role played by university mathematicians and the structure provided by the quarterly meetings, as well as the long history of camaraderie among the faculty, a new mind-set of openness and sharing was fostered. Teachers appeared eager for opportunities to communicate mathematics to each other as a means of strengthening their own conceptual and methodological understandings. One local administrator described the value of this transformation, noting that the project provided teachers with an environment in which they "have a chance to try new ideas and talk to each other about their successes and failures—a whole new kind of peer coaching for us." A Desert View teacher elaborated,

> When we first got together as a project group … there were definitely some barriers between us as far as being afraid to criticize or damage egos. Through the process of editing projects for each other, we have learned each other's worth. And it's helped level everybody out to an even playing field where they feel more comfortable communicating back and forth …. Getting over the barrier of having done something dramatically different in the classroom and being successful at it has made everyone believe that it's not that tough to try new things. Now it's almost expected that you are going to try new things, see what happens, and not be afraid to report back to your peers about what worked or what didn't work.

Creating the climate for this supportive and risk-taking behavior was not simply a matter of fostering collegiality among Desert View's teachers. Although the mathematics department benefited from a long tradition of social cohesiveness, several factors appeared central to the teachers' successful collaborative stance with project-related activities. First was the sense that teachers were being validated as professionals through their work with the university. "They saw that we were taking their knowledge and experience seriously," noted one professor. "We came into their classrooms, not to say, 'You should do this and this and this,' but to observe and learn what worked." Correspondingly, at the district level, members of the department noted the willingness of administrators to send teachers to conferences and national meetings without "checking up" on them afterwards—confirmation, in the teachers' eyes, of their own competence and professional status.

This sense of validation among teachers manifested itself in a new form of "professional collegiality" that, in the words of one department member, "has the faculty working toward mathematical goals at a more efficient and sophisticated level." In conjunction with this transformation, the very structure of the project approach encouraged a new emphasis on collaborative, long-term planning.

> Another nice thing about the projects is that when we plan, we have to plan long-term. When we [teachers] work out the problems and projects among ourselves first, it gives us the opportunity to find out where the students are going to get messed up, to anticipate what we are going to have to help them with, and to determine what the students are going to need to figure it out on their own or amongst themselves.

The climate at Desert View appeared receptive to these new collaborative thrusts, due in part to the mathematics department's existing policy that ensured equitable teaching loads. In contrast to a seniority-based system for determining teaching loads, Desert View's department required only that each teacher instruct one remedial or lower level class. Beyond that, each teacher received his or her first choice of a class—or at least a second choice with a promise of getting the first choice the following year. Remaining classes were assigned according to scheduling needs and, to the extent possible, continued teacher preference. One first-year teacher, for example, shared her delight with a teaching load that included two sections each of Algebra I and Algebra II and one remedial "math applications" class. Reflecting the policy of equitable loads regardless of seniority, the same year, one of the department cochairs taught a remedial algebra class and two sections apiece of Algebra II and geometry.

Creating a New Category of Successful Students

Enhanced communication among the teachers helped to foster what one teacher described as the program's "humanizing effect" on the students, resulting in their willingness and ability to communicate more freely with teachers. "The projects," observed one district administrator, "have given teachers a relationship with the kids they never had before because the projects demand a new type of interaction between teacher and student." In our efforts to "unpack" the meanings that underlie this new type of interaction, we sought input from two of the teachers who initiated the collaborative program with the university. The teachers' responses focused on two issues, teacher flexibility and new expectations regarding student interest and achievement. Characteristic of teachers' understanding of flexibility and changing expectations were the following comments:

> One thing I have found that has changed since I have been doing projects versus when I didn't do projects is that I am much more flexible. I tend to be more aware, on a daily basis, of who is understanding what. I think it is because the kids are freer to communicate something that they do not understand. And I don't feel like I am locked into some rigid schedule where I am covering this today no matter what, and tomorrow we are going to be here, and then Tuesday we are going to have an exam. I really feel my flexibility is a direct result of more communication between me and the students.
>
> The key issue is that, as a result of teachers' dabbling with projects in the past two years, they now have an increased appreciation that students can do something beyond what they thought they could. A lot of students that are generally interested in learning and have a good attitude about learning never saw math as offering anything stimulating to think about. Many of those students find that they are successful in the projects program partly due to the group interaction. I think that helps spur interest because they now have an opportunity to express themselves creatively, whereas before in the classwork, all they could give was an answer that was right or wrong. This is what I think is the humanizing effect that lets these kids open up, that has allowed a category of kids to be successful who weren't successful before.

The teachers at Desert View High School consistently expressed their commitment to the notion that mathematics should serve the needs of all students and not just those who are college-bound. However, the constraints, structural and otherwise, that hindered the teachers' ability to enact this vision as they saw fit were apparent. One teacher, for example, noted the problems inherent in a course sequence that, despite the best of teacher intentions, continued to gloss over the needs of a substantial portion of the student body:

> The "algebra, geometry, trigonometry" sequence that we're locked into only seems to serve science, prescience, preengineering, and premedical majors and doesn't do anything for about 60% of the general student population who are interested in what math can do. The program as set up is oriented toward preparing students for calculus. I teach some of the remedial algebra classes to kids for whom everything we're doing is really of no practical use. I know from the few times that I have done something that was more open-ended or calculus-oriented that the kids are naturally curious about numbers. But we manage to deaden curiosity by the time they reach this level. Certain barriers, like an algebra barrier, contribute to this—you either succeed with traditional algebra or that's the end of your math career.

Nearly every teacher with whom we spoke raised the corresponding concern of student absenteeism, noting especially the compromising effects of family socialization and peer expectations on students' commitment to schooling. Although they recognized considerable in-group variation among the large number of Hispanic students who attended Desert View High School, generalized perceptions of minority student achievement persisted among the faculty:

> Absenteeism is probably the worst problem we have. It is frustrating and there is nothing we can do. And a course like algebra is so layered that if you're missing a layer you can't continue. The kids don't understand. Some of the Hispanic students believe that being at home and taking care of a younger brother is more important than going to school. Others view their role as a gang member as a top priority. Some of them do not see the value of school.

Such general frustrations—whether rooted in cultural or socioeconomic differences or in the structure of the school itself—did not translate into a passive or fatalistic acceptance of poor student performance. To the contrary, most teachers assumed a proactive stance that mirrored the risk-taking, "can-do" attitude that underlay their adoption and ongoing adaptation of the project approach. Few mathematics teachers were fluent in Spanish, but at least two had taken steps to receive further training in ESL and others actively sought student teachers and aides who were Hispanic or at least bilingual in Spanish. The high school principal maintained an active counseling and mediation group with gang leaders within the school, the effects of which were evident in the cooperative group environment of the mathematics classrooms. As the teacher of a technical math class noted during one of our observations, "Many of these students belong to rival gangs, but in my class they work together in groups."

The question of how to tailor a course for students who had never experienced success in mathematics was of utmost concern to Desert View's teachers. At the time of our initial visit, projects represented only a small portion of overall mathematics instruction and had not yet been incorporated in any of the remedial classes. This relatively minor level of emphasis belied the more tacit influence of the project approach on teachers' desire and ability to implement reform-minded changes in *all* levels of mathematics classes. As observed during our subsequent visit, the increasing use of multistep, nontraditional mathematical activities (i.e., projects or modified versions of projects) in both low- and high-level classes contributed to a general vision of mathematics as an integrative experience in which students were encouraged to draw on multiple strategies in a cooperative, hands-on format to solve real-life problems. However, practical concerns sometimes tempered this vision, as one teacher explained.

> I structure [my remedial class] differently than my other classes because these are the kids that tend to have a high rate of absenteeism. They have a hard time being responsible enough to get their homework back to class every day, show up with a pencil, that sort of thing. They need short-term kinds of goals. I try to make sure they don't have a lot of work to do outside of class because many of these kids don't have a home life that is conducive for studying. The time I spend at the board actually giving them new ideas and all for their notes is fifteen minutes, max. They get the context down, we practice together, do some examples together, and then they are off doing that day's activity for class.

Neither this teacher nor her colleagues with whom we spoke seemed totally satisfied with such rationalization of the instructional process. At the same time, they continued to celebrate whatever small gains were made by students and expressed pride in their own persistent efforts to foster students' success.

> I think our whole department is proud of the fact that we've had these kids who have never had any success in math—I mean they have flunked course after course after course—and we have intentionally looked for ways to give them some kind of success in math. If they just learn to do their checkbooks, hey, every time they do it, we're a success!

Doing Mathematics

"Doing mathematics" can be interpreted in different ways. In the past, doing mathematics meant working a number of similar problems until a rule or procedure was memorized. Now, by *doing mathematics* many mean an active construction process of learning mathematics (Lampert, 1990; NCTM, 1989). However, as Lampert (1990) pointed out, there is a need for pedagogical change in order to foster a different "doing" of mathematics and the linking of "knowing" with "doing":

> [C]hanging students' ideas about what it means to know and do mathematics was in part a matter of creating a social situation that worked according to rules different from those that ordinarily pertain in classrooms, and in part respectfully challenging their assumptions about what knowing mathematics entails. (p. 58)

Desert View teachers' use of projects reflected—to some degree, incidentally on their part—the interpretation of "doing mathematics" presented in the *Curriculum and Evaluation Standards for School Mathematics* (NCTM, 1989). Specifically, teachers were facilitators to the learning process, which frequently took place in cooperative group settings where students explored challenging and realistic problems. There was less emphasis on amassing more and more mathematical facts, and "doing" replaced "knowing that" as the criterion for understanding mathematics. According to the *Standards*,

> ..."knowing" mathematics is "doing" mathematics. A person gathers, discovers, or creates knowledge in the course of some activity having a purpose. This active process is different from mastering concepts and procedures. We do not assert that informational knowledge has no value, only that its value lies in the extent to which it is useful in the course of some purposeful activity. It is clear that the fundamental concepts and procedures from some branches of mathematics should be known by all students; established concepts and procedures can be relied on as fixed variables in a setting in which other variables may be unknown. But instruction should persistently emphasize "doing" rather than "knowing that". (p. 7)

Desert View teachers attributed to their work with projects an increased concern for what mathematical content students should learn as well as for the process of communication. The motivation for developing particular projects often stemmed from student needs—especially with regard to concepts that were difficult for many students. The teachers' concern for content and communication translated into a twofold emphasis on real-life applications of mathematical understanding and the development of writing skills in mathematics. With respect to real-life applications, teachers spoke frequently of their efforts to relate projects to everyday life and to connect mathematics to something else in the students' lives. One teacher explained,

> I don't like to give a project that just says, "Crunch all these numbers out." They get enough of that from the textbook. So, for example, I gave students a project on building a fence around the athletic field. They were given some limitations, and they had to give me the dimensions that would minimize the cost of the fence. I like to give a project that, maybe when they read it the first couple of times, they don't see anything mathematical there, but mathematics is the way they solve it.

In a similar fashion, writing was viewed by teachers as an effective vehicle for building students' awareness of the distinction between completing a set of mathematical tasks and understanding the concepts underlying such tasks. During an interview, one teacher raised the following issue: "How many times do you have a student tell you, 'I did this, but I don't know, I really don't understand how it happened'?" Another teacher applied the above remark to the program's focus on writing:

> The writing that is tied to the projects is a big item. I never realized—maybe because I was never made to write in a math class—what a difference it can make for students to see not only if they know how to do a problem but also how to explain the problem. I think the students are beginning to see some purpose to math and the difference between being able to do math and understanding math.

Desert View teachers expressed satisfaction with the high levels of reasoning evidenced in writings of students who were involved with projects. For example, an Algebra II teacher proudly shared several completed projects in which students had to generalize graphing strategies for translations, reflections, and dilations. Instead of memorizing rules on graphing techniques, the students in this class had searched for patterns through an inductive process, examining the effects of altering functional equations in a variety of ways. Discovering these generalizations was the easy part. Writing their findings so that others would understand them required deeper mathematical reflections and reasoning.

Students were no longer motivated to get "quick" answers to problems. Instead, they challenged each other with their own ideas and alternate solutions until a group consensus was achieved. This resulted in increased levels of pride in their work. For example, one teacher indicated that as a result of more pride in accomplishment, students held on to their completed projects and significantly fewer assignments ended up in the trash can, having been thrown away by the students or the teacher.

> In classes that I was using the traditional methods, year after year, as soon as the paper has been recorded in some way, high piles of math papers just end up in the garbage, and I used to feel bad about that, especially when I first started teaching—that they would find they didn't need these certain notes or this certain test .... Then I realized that there was no reason to keep them because most of them [student papers], if you went back to look at them a year later, would be very irrelevant ... There was very little of the students, not only intellectually but emotionally, in these pieces of work; there was no attachment to the work at all .... Now I've got a file full of work from students from last semester. I need to make room in their folders to add additional work this semester, but I can't throw out these priceless pieces of work.

During our visits to Desert View High, we were able to observe projects at various stages of development; in some cases, we observed a project activity as it was introduced and in subsequent weeks examined students' finished products, which had been mailed to us. We were particularly interested in looking beyond the changes in teaching and learning that were occurring in classrooms in order to get a sense of the projects' mathematical content as well as the level and logic of student reasoning involved in completing it. In order to take this "deeper" look, we listened carefully to what students were saying to each other as they worked together in groups, asked mathematically-specific questions during the student interviews and classroom observations, and examined students' completed projects.

In the following descriptive segment, we explore how, and to what effect, the "doing" of mathematics was manifested in one of Desert View's geometry classes through the use of a project. In the second part of this section, we describe the impact of project-related activity on students who had previously experienced little or no success in mathematics.

Constructing student knowledge through projects. The teacher introduced the new project, entitled "A 'Regular' Dilemma," by distributing a tetrahedron model made of colored paper to each of her geometry students. "There's a surprise inside," she announced. Students began to share the scissors that had been distributed, and to snip through the scotch tape holding the three-dimensional model together. "Now take out the paper and read it carefully." Students, clustered in groups of four, began to extract a tightly folded sheet of paper from their tetrahedron models.

The paper instructed them to form a fictitious floor company that specialized in using tiles in the shapes of regular polygons. The students' first customer was to be a soccer fan who wanted the design of a soccer ball on the floor of his recreation room. The teacher asked, "If a soccer ball is taken apart and laid flat, could we tile a floor with the pattern?" The question served as a "motivating context" for the inquiry that followed.

"A 'Regular' Dilemma" was rich in geometric content. It enabled students to use tesselations to investigate in a hands-on fashion various polyhedra and patterns involving regular polygons. It challenged them to (a) work with regular polygons of up to 12 sides; (b) look for the relationship between the number of nonintersecting diagonals each polygon contained and the number of nonintersecting triangles; (c) find the sum of the measures of the interior angles of various polygons and generalize their findings to any n-sided polygon; (d) apply their findings by determining the feasibility of the soccer ball floor pattern; and (e) explore other possible tile patterns using one or more regular polygons. Students were encouraged to extend their applications to individually created problems and story lines as well.

On this first day with the project, some of the groups spent time determining individual responsibilities such as writing the story line, doing the graphic drawings, formulating the mathematics needed for the task, and editing the final project. Other groups immediately began working on the mathematical tasks.

By the second day, most groups had discovered that the soccer ball pattern was not suitable as a floor pattern. Student responses to our questions about their work revealed their developing understanding of the mathematics needed to determine the measure of interior angles of regular polygons. The following dialogue between one of the investigators and one student helps to illustrate the students' line of reasoning.

I: How do you know the soccer ball pattern will not work?

S: [In response, the student made a sketch, Figure 4.1, of two pentagons adjoining a hexagon at a common vertex.] See (pointing her pencil to the common vertex), there is some left over.

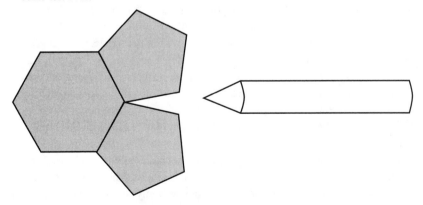

Figure 4.1. Student sketch showing that a soccer-ball pattern will not tessellate

I: Some left over? I don't understand.

S: Well, we figured that these three angles don't quite add up to 360 degrees.

I: Why is that?

S: Oh, you want me to explain how we got the angles?

I: Yes.

S: Well, in a five-sided figure (again pointing to the pentagon), there are three triangles. (See Figure 4.2.)

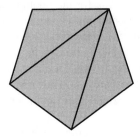

Figure 4.2. Student sketch showing that a pentagon consists of three triangles

So that's three times 180, or 540. Since there are five angles (pointing to successive angles around the pentagon), that's 108 for each angle. [She repeated the same process for a hexagon and arrived at interior angles of 120° each.] Now you have two of these angles and one of these (pointing to the figure), so that's 120 plus 120 plus 108— which is twelve less than 360.

We were curious about whether the students in this geometry class had studied this material at some earlier level. A few students remembered learning the expression $(n - 2)180$ in the eighth grade, but they felt that now they really understood where it came from and how it could be used.

Later, as we examined finished projects, we were able to determine in greater detail the logic by which students arrived at their conclusions. Computer-generated renderings of regular polygons traced the progression of student understanding from nonintersecting diagonals in various polygons to the final generalization needed to complete the mathematical requirements of the project. Excerpts from a completed project (see Figure 4.3) illustrate one student's mathematical reasoning.

We gained additional insights into students' problem-solving abilities through informal questioning as students worked in small groups. The following conversation provides glimpses of students' abilities to apply and communicate their mathematical understandings.

To begin researching for my first job as a subcontractor for Mitla Construction Company, I used a straight edge and neatly drew seven regular polygons—a triangle, square, pentagon, hexagon, octagon, decagon, and dodecagon. Using the definition of diagonal (segment joining two nonconsecutive vertices of a polygon), I concluded that a triangle does not have a diagonal because its vertices are consecutive.

Using the drawings of the seven regular polygons [see example of the pentagon below], I drew in all the nonintersecting diagonals and found that triangles were formed within each polygon. The sum of the measures of the three interior angles of a triangle is 180°. I found that if you multiply the number of triangles within the regular polygon by 180, you can find the sum of the measures of the interior angles of each regular polygon. The sum of the measures of the interior angles of a square is 360°; for a pentagon it is 540°, for a hexagon 720°, for an octagon 1080°, for a decagon 1440°, and for a dodecagon 1800°.

The pattern noticed relating the number of sides, number of interior angles, and the number of triangles formed within the polygon is the number of lines joining the vertices is equal to the number of sides. Also, it was noticed that the number of sides is two more than the number of triangles formed within the regular polygon. Therefore, in the formula $n - 2$, the variable, n, represents the number of sides of a regular polygon and would be used to find the number of triangles formed within any regular polygon.

Taking what I have learned, I decided that a formula was needed to determine the sum of the measures of the interior angles of any polygon. I took the first formula, finding the number of triangles in a regular polygon, and multiplied it with the sum of the measures of a triangle and developed this formula: $(n - 2)180°$. Now, simply insert the number of sides of the regular polygon for the variable, n, subtract 2, and then multiply by 180 to find the answer.

Now, since I know how to find the sum of the measures of the interior angles of any regular polygon, I need a formula to find the measure of ONE angle in that regular polygon. I took the formula

$$\frac{(n - 2)180°}{n}$$

The n represents the number of sides, which is equal to the number of angles in that regular polygon. This makes it possible to find one individual angle.

Shayley's Tile Company was asked by Mr. Brady, a soccer fan, to tile his game room in the pattern of a soccer ball. However, I found this would not be possible. The problem is that where the two pentagons and one hexagon meet, a small gap remains. The measurements of the three regular polygons are 120°, 120°, and 108°. The sum of these numbers equals 348°. In order for the floor to be properly tiled, the sum must equal 360°. When 348° is subtracted from 360°, the difference is 12°. This 12 shows the angle of gap between the combination of these three polygons. So Mr. Brady was disappointed and decided to carpet his floor.

[The student displayed the following sketch with his write-up to show how he arrived at an interior angle of 108° for a regular pentagon.]

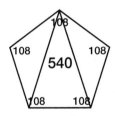

Figure 4.3. Portion of a completed project—a "regular" dilemma [as displayed by student]

I: What is your story line?

S1: Well, there's a good guy that's—

S2: —in a cave and captured. He's in an enemy's kingdom, and they have them make shields.

S1: What they're trying to do, they're trying to find out what kind of shields they need. 'Cause they try to fit the shields together like that ... We're thinking about making the tile thing but with shields so you can put above your head, like to be protective when they're engaged in battle.

I: So the shields will have the patterns of the polygons?

SS: Yeah.

I: Have you figured out which patterns will work?

S3: Yeah, we're working on how to draw them.

S4: We've found octagon and squares.

I: Have you figured out some sort of mathematical test to find out when things work or not?

S3: Yup. See, together (student pointing to vertices of adjacent angles of three polygons) this will add to 360°.

I: So, that's your test?

S2: The problem with the soccer ball thing was ... this is 120, and this is 120, and this is only 108 (pointing to another sketch of hexagons and pentagons, Figure 4.1). So you had about twelve degrees left over, and you had a little bit of a gap here.

I: So you can't use this pattern, then?

S2: Yeah.

Further exploration led to the general case that the total number of degrees in the angles of an *n*-sided polygon was $(n - 2)180°$ and that the total number of degrees in angles about a common vertex would need to total 360° if there were to be no gaps. Some students recognized this generalization as a rule that had been previously presented in another mathematics class, and they now applied this generalization within the context of the project.

Project-related efforts across ability levels. According to Desert View's mathematics teachers, doing mathematics through project-related activities had a profound impact on students who were considered to be at risk. The teacher of a "geometry essentials" class explained that the students in his class have "severe, diverse problems both at home and at school that hinder learning in a traditional school setting." He explained that one of his most important tasks is "to convince my students that it's okay to fail, that failure often leads to success. Even the most successful hitters in baseball fail at least two out of three times."

> See that girl in the orange blouse, the one talking to other members of her group? She never said a word until two weeks ago, not even to me. I consider her one of my biggest successes. The projects make mathematics interesting for these kids.

The group to which this teacher refers was working on a project to determine the size of a giant. Each student had received a photocopy of a giant's hand on a sheet of paper. From that, they had to estimate the height and weight of the giant using proportionality based on actual measurements of themselves. In response to a request for help, the teacher provided the following explanation:

> Let's dream awhile. For example, I received word that a giant has come to the village and taken my child. I'm upset and want to rescue her. Think, now! If you haven't seen the giant, wouldn't it be a good idea to get an understanding of what you're up against? So let's prepare ourselves with some information about this giant. What information do you think you need?

The teacher then made the suggestion that they could determine the area of a hand by using a graphing grid. He followed this up by showing them how they might use a graphing grid.

In a math applications class, we observed groups of students in a lab setting gathering data relating to small boxes of M & M's (which were given to each student and were to be eaten after the data was gathered). This activity would be described as a "mini-project" because it would be completed in a couple of days.

This was a class considered by teachers at Desert View to be a difficult group of students. Most had failed other mathematics courses, and many had experienced discipline problems in school. Each student was to weigh five of the candies and then determine the average weight of an M & M. This information was to be used to predict how many M & M's are in an average-sized box of M & M's. Students were also asked to look for frequency patterns of the various colors of M & M's. On the basis of these frequency tables, students were to predict what percentage of M & M's in a typical box would be green, brown, tan, or yellow. Individual results were to be compared with others in the group. Then a composite result would be derived by each group.

These activities were accompanied by a great deal of talking in each of the groups. One of the three groups consisted of one girl and four boys. This group started out slowly until the girl in the group took the lead and directed the activity. She really "got after" the members of her group, and they were soon following her orders.

The teacher circulated from group to group, checking on progress and answering questions. "It keeps bouncing, Miss," said the boy weighing the M & M's as he stood by the scale, moving his index finger up and down to imitate the motion of the balance scales. The teacher walked up and assisted him. Several students joked with the boy in a good-natured way, as the teacher demonstrated how to steady the balance scale.

At this point, we observed a group of four boys. This was a quiet group. One student was very precise in the weighing of his M & M's and his recording of data. Students in this group were collaborating, but were more serious than the other groups as they quietly worked together. Later, as students in this group compared results of the M & M's (i.e., color frequencies and weights), the boy who had meticulously weighed the M & M's turned out to have percentages that had the biggest divergence from those of the group. This boy seemed very disappointed that his "care" in doing the task had not placed him closer to the group norm.

This mini-project was not a real-world example of mathematics application. Davis (1964) discounted the need for mathematical experiences to be based on concrete experience. "The children do think more creatively when the ideas are *meaningful,* but the *meanings* do not have to be *concrete!*" (p. 6). We observed that the M & M experience was "meaningful" to these students, and it contained a number of mathematical concepts. This mini-project was rich in notions about statistics and probability (e.g., the collection and analysis of data; determining frequencies and percentages of M & M colors, the probability of getting each color in a sample, deviation from norms). It also demonstrated a need for important mathematics skills such as working with percentages and fractions (i.e., part of a whole) and estimating.

The examples above are a sampling of what we observed during the 6 days we visited Desert View High School. The pedagogy in mathematics classes was in transition. Instances of "traditional" teaching practices were still taking place, with teachers imparting information to students, who passively received it. But increasingly, teachers were offering their students opportunities for learning mathematics by doing mathematics.

PARTICIPATING IN A MATHEMATICAL VISION:
LEARNING FROM DESERT VIEW

This case study presents the promising image of an instructional innovation that directly and indirectly served to catalyze and sustain mathematics reform in a traditional high school setting. Though our account describes specific events bounded by time and circumstance, our intent is to underscore the broad dynamics of ownership, leadership, and risk taking in the process of change. In addition, we have highlighted classroom practices that, to varying degrees, reflect the aims of the NCTM *Standards.* Notable for their absence in Desert View's case are at least two factors commonly observed in other R^3M study sites, namely, a "visionary leader" to instigate and guide change efforts and a neatly articulated vision to which all participants subscribe equally and explicitly. To the contrary and, ultimately, to Desert View's credit, project-driven change in the high school was neither dictated nor uniform.

As a potential or partial model for mathematics reform, Desert View serves to alert us to the functional advantage of individual variation in planned educational change. Desert View's mathematics teachers operated less out of a perceived need to fit into one particular vision of instruction than on the basis of their own attempts to define a range of acceptable practices that reflected their developing understanding of the project approach. Set within a loose but effective structure of shared leadership among a core group of teachers, this individual autonomy offered members of the mathematics department the opportunity to shape a practical perspective informed by selective and gradual identification with project-driven reform. The teachers' positive orientation toward each other's diverse needs and strengths further enabled them to proceed at their own pace and on their own pathways of change.

What we encountered during our visits to Desert View was a mathematical vision that was being creatively acted on but not consciously or consistently articulated by the teachers. Desert View teachers were clearly acting on their beliefs about mathematics reform and then, in effect, were letting their actions speak for their individual interpretations of that reform. Again, to Desert View's credit, thoughtful *adaptation* of the project approach superseded unreserved *adoption* of it. Enacted in an environment already noted for its supportive and risk-taking nature, this process of gradual adaptation enabled teachers to draw on a common vision of mathematics without being forced to make an "all or nothing" commitment, at least initially.

Herein lies a key insight from Desert View's experience. Those teachers committed early to the project approach maintained a patient and flexible stance with regard to the other teachers—acknowledging their colleagues' early fears, openly sharing their own successes and failures, and creating channels of support so others could take risks *when they were ready to do so*. Consistent with this image was a district administrator's characterization of teachers who "are moving in the direction appropriate to the program":

> They are asking themselves a lot of questions about what they are doing and they are trying things they never did before—taking risks in the classrooms in terms of the way they teach … [T]hey have come to realize that everything they do doesn't have to succeed. They have talked about their successes and failures. When one [teacher] has not been as successful as the other, that successful teacher has said, "Well, I did it this way"—giving the second teacher a chance to think, "Well, I might try that again and tinker with it a little."

A comfort with risk taking and the freedom to fail cannot be mandated, but they can be nurtured. In Desert View's case, a policy of equitable teaching loads, a sense of professionalism enhanced by support and respect from university mathematicians, and the provision of time and resources for ongoing collaboration with colleagues enabled teachers to come to terms with a mathematical vision in a deliberate and unthreatening manner. Desert View's experience with project-driven reform demonstrates clearly how the two-way exchange of innovative ideas between public school practitioners and university mathematicians can be facilitated. Perhaps the most significant and replicable aspects of Desert View's efforts were the opportunities created for risk taking and long-term collaborative planning among teachers. Moreover, it is clear that such opportunities were optimally fostered in a climate that valued teacher ownership and autonomy. Even so, different individuals in the Desert View setting assumed significant roles at different times, depending on which skills (e.g., organizational, consensus building, problem solving) the situation demanded—contributing, in effect, to a shared leadership in reform efforts.

Coda: A Glance Back at the Standards

Although our discussion has not centered on the role of the NCTM *Standards* in Desert View's reform efforts, we conclude with several comments and questions that do. Both teachers at Desert View and mathematicians at the university agreed that the projects were not originally intended to address the *Standards*. They also noted that the teachers were much more aware of the *Standards* than the professors and were much more willing to approach issues in the *Standards*, such as problems with statistics. Interviews with teachers revealed differing perceptions of the impact of the *Standards* on the overall mathematical vision of the department. One teacher commented,

When the *Standards* first became available to us in published form, people had a lot of anxiety and kind of set them aside and said, "We'll see what we can do but we don't expect too much." It seems like it was not until a year ago [1992], after we had a successful program with writing and projects, that we started looking back at the *Standards*. We realized that our program was aiming our students toward many of the *Standards*, and it was an affirmation that we must be doing the right thing ... but we never thought of our program as a way of approaching the *Standards*.

Like other sites in the R^{3}M study, Desert View illustrated the validating, rather than generative, influence of the NCTM *Standards*. Teachers interpreted the *Standards* as applicable in retrospect; they operationalized and affirmed the *Standards'* underlying concepts only after teaching mathematics through their own constructed projects. One teacher explained the NSF project's facilitative effect on his understanding of the *Standards*:

> A lot of the points in the *Standards*—problem solving, cooperative learning, communication—were at first overwhelming, even if you agreed with them. Now that I have implemented many of them in class through this project, I can see that they are not too difficult.

Perhaps, as another teacher advised, it was not productive to focus on the horse-and-cart question regarding the projects and the *Standards*. What mattered, she asserted, was that the goals and results of both were similar and mutually reinforcing: asking students to build models, deal with real-world phenomena, communicate mathematically in both written and oral form, and engage in mathematics as an integrative experience in which concepts are not isolated from each other and students are encouraged to draw on multiple strategies in a collaborative fashion.

Desert View's case highlights what may be a missing part of the *Standards*, namely, how to help teachers cope with cultural and social diversity in a meaningful manner, particularly when students' cultural expectations conflict with those of White, mainstream America. This issue encompasses the way mathematics is viewed in the home, early number learning, verbal or written emphasis in the home, parents' mathematical knowledge base, and the overall value placed on formal education. A problem-solving approach such as the one adopted by Desert View in the teaching of mathematics can contribute significantly to overcoming traditional barriers of obscure and often nonsensical tasks students are asked to complete. Further progress toward the creation of new categories of successful mathematics students demands that we look at the *Standards* in light of increasing diversity among students and between students and teachers. It is within culturally and socially diverse settings such as Desert View that the role of the mathematics teacher can assume special significance—and through which reform-minded educators can craft new and necessary insights into the NCTM *Standards*.

Chapter 5

Evolutionary Reform at East Collins

Joanna O. Masingila, Patricia P. Tinto, and *Loren Johnson*

NOTICEABLE CHANGES

Located just an hour from a major airport and city, East Collins High School is accessible by a twisting two-lane road that leads past small farms, horse ranches, and new housing developments. The school, a single-story complex, is a series of wings surrounded by large parking lots. The building opened in September 1988 with fewer than 800 students in Grades 9–12 and now has nearly 1500 students, comprising 72% White students and 28% Black students. The students come from a wide economic range, from wealthy families to those living in subsidized housing. An administrator at the school noted that the student population had changed from a small, intimate group that had been together since elementary school to a large group that includes many students new to the district.

Besides changes in the size of the student population, there also appear to have been noticeable changes in the last 3 years in the mathematics education the school offers. Consider the following "snapshot" from an Algebra I class. The students are in the computer classroom entering, editing, and answering questions about the following BASIC program.

```
10 INPUT X
20 IF X > 5 THEN GOTO 50
30 PRINT X; "IS NOT IN THE SOLUTION SET"
40 GOTO 60
50 PRINT X; "IS IN THE SOLUTION SET"
60 END
```

Students are working in pairs and communicating mathematically. One pair enters some input values and then discusses what the program does.

S1: We got "12 is in the set; –4.5 is not in the set."

S2: So the program decides if a number is negative or not.

S1: Or how about if a number is a whole number?

S2: Let's try something else.

S1: Put in a 3.

S2: Oh, look at this line (points to line 20). It only puts this if *x* is greater than five.

S1: So this tests whether a number is greater than five or less than five.

S2: Change line 20 and see what happens.

Conjecturing, reasoning, and communicating are now commonplace in mathematics classrooms at East Collins High School. However, this has not occurred without concerted effort on the part of many persons involved with the mathematics program there. This effort has taken the form of a project called the Perfect Situation for Mathematics Learning (PSML).

A PHILOSOPHY FOR MATHEMATICS TEACHING AND LEARNING

The Perfect Situation for Mathematics Learning

The Perfect Situation for Mathematics Learning came to life as a partnership between the State Power Company, Southern Mills, the Collins County Schools, and the State Department of Education. Planning for the PSML project began in January 1991 and represented the combined efforts of teachers, administrators, and mathematics professionals working with business leaders to create a different mathematics learning environment. PSML was created with assistance from the State Coalition for Excellence in Mathematics Education, and the PSML team—which consisted of two mathematics teachers, the district mathematics coordinator, the associate principal, the superintendent, an industry representative, and several college mathematics professors from the Coalition—envisioned the project as being firmly grounded in the *Curriculum and Evaluation Standards for School Mathematics* (National Council of Teachers of Mathematics, 1989).

The main goals of PSML were for all students to (a) become autonomous learners, (b) have a common foundation in algebra, (c) develop better attitudes toward mathematics, and (d) persist longer in mathematics. A brochure describing PSML notes,

> In a PSML classroom, students learn to solve problems through a hands-on approach, using both modern technology and everyday materials. Abstract principles are made concrete as students conduct lab experiments—generating data, posing problems, and employing mathematics principles to arrive at solutions. Computers, graphing calculators, journals, and teamwork strategies are used to integrate mathematics concepts with technology and verbal skills. Reading and communication skills are stressed in completing individual and group tasks, much like assignments in the workplace. Teachers work with students on *how* to solve problems—rather than dictating solutions and assigning repetitive sets of calculations.

What drove the development of PSML was the recognition by mathematics teachers that a large number of East Collins students were choosing to enroll in nonalgebra mathematics courses in their ninth-grade year and that these students were not as challenged mathematically as they might be. The chair of the mathematics department indicated a readiness for mathematics reform at East Collins High School (ECHS): "We just felt as a math department that there was a better way, and so that feeling made us very receptive. So when we had this opportunity, we just jumped at it. It was great timing."

The associate principal and the district mathematics supervisor also recognized the need to have more students taking algebra and stressed the need to bring relevant mathematics to the district's students. Additionally, recent state requirements mandated that all students graduating from high school complete algebra. However, recognizing a need and being able to implement change are very different processes within public schools. The catalyst for change came in the form of a business partnership. Implementation of the project began at ECHS and East Collins Middle School in the fall of 1991 with Algebra I and prealgebra classes. The mathematics chair explained the shift in power from the original team as the implementation phase began its early stages:

> After we started last year, the committee voted to make [our associate principal] the head of the committee, and the money was put into an account in our school system, and I handled the disbursement with her approval. So we [the two lead teachers in the project, the associate principal, and the county mathematics coordinator] now pretty much make the decisions. It's up and running and the other folks have said, "Well, it's yours now." And they're still there, and we still meet, and I like to hear what they have to say, and they have real good ideas. But basically we're making the decisions.

PSML expanded from including Algebra I in 1991–92 to including Algebra II during the 1992–93 school year and Euclidean geometry in the 1993–94 school year. Funding from State Power helped to provide for the purchase of some textbooks, supplementary materials, manipulatives, a computer for each mathematics classroom, and summer staff development meetings during the first 2 years of the project.

The Future of PSML

An PSML brochure describes the project as an innovative *process* for learning mathematics. It is not a new educational product with a limited kit of materials, but rather a constantly evolving model able to accommodate new teaching insights as they emerge. The mathematics department chair at ECHS corroborated this view:

> PSML is evolving. As it evolves, I'm learning constantly; everywhere I go, everyone that I talk to, I learn something from them, and the *Standards* are part of that, so it's kind of a mind set, I guess. We're looking for these sorts of things that are the *Standards*.

The chair also shared his personal investment in the PSML project.

> I really feel strongly about this .... I hoped that by accepting the responsibility of chairing the math department I could offer leadership and guidance to expand PSML. A lot of times a good idea comes up and everybody gets excited and for two or three years it is there, and then it just kind of fades away. It's an overall philosophy of change and things that need to be done in the math classes from K–12. I'd like to see that throughout our whole system ... I know our mathematics coordinator would like to see that. It goes way beyond with what we're doing here in our school.

Two teachers took the lead in implementing the project in ECHS. Others in the department recognized the importance of these two who pioneered PSML.

> I think that we would not have been nearly as successful as we have been without them knowing where they wanted to go, but they didn't have to create an entire thing. They had a base to build on, and they brought in all their manipulatives, the leadership of State Power ... I think this is all going to get better too, as we all get some experience with the materials we have and the means that we have for different avenues, and we can find some time to talk to each other about it.

One of the lead teachers believes that as PSML expands, "it's going to start changing and growing." She also feels that as a result of the materials and the technology available in the new program, teachers will reflect on and rethink their classroom practice.

> It's creeping around the building in other departments. Yeah, it's parading around the corner into the science department. The English department is going "um." You can see it creeping around the building and with that, it starts to spread out outside the building. We are hoping to continue to grow up and down.

Another teacher expressed her hopes that the district won't "hold back" but will allow PSML to "trickle on down to the elementary level." Other teachers noted that PSML is moving up to the higher level courses faster than was originally planned. One teacher spoke about using some PSML laboratory ideas in a trigonometry class and the students having "loved it." The teacher noted,

> Most of them have not been in this program [PSML]. They're staying just ahead and they had never done that, so they really enjoyed it. They wanted to do more.

After the first year of the project, the associate principal indicated that PSML was growing stronger but believed the program still had some vulnerability.

> I feel very confident that the program could stand alone. I guess that it's just that the two of them [the lead teachers] have done such a good job of getting it started that I think it's been better for us as a result. And it's important for all of us to stay involved with PSML; it's not something you can leave behind for the next teacher just to pick up, you've got to do this kind of thing, all this talking, all of this explaining, all of the showing. That's the scary part. It is dependent on people and it is something not written down. I'm not sure we could ever put in written form what PSML is.

After the second year, however, the associate principal felt that PSML was solidly in place.

> One hundred percent [of the mathematics teachers] agree with the philosophy because it is not an outrageous philosophy. I would say that all of them are so indoctrinated—and I use that word openly—so indoctrinated to the PSML philosophy that they would have a hard time saying where they differ …. What's happened is that because the math teachers have worked together, trained together, had staff development together, they talk together. It's been almost a natural progression.

The assistant superintendent also felt that PSML had passed its major hurdle:

> I would have worried last year about the program if for some reason we had key people going somewhere else. But this year, now with other people involved, I don't think it's nearly as big a deal.

In fact, after the second year of the project, one of the lead teachers left ECHS because her husband was transferred to another job location. We observed that the evolving process of PSML continued because more teachers had become heavily involved and carried on the implementation of the project. One teacher explained, "I don't think that there is any teacher that I have seen, any teacher that is involved at this moment who doesn't want to be, and that's saying a lot. Everybody that's in it wants to be doing it and wants very badly to be doing it."

Furthermore, the chair and the district mathematics coordinator reported to us that the other high school in the county was interested in what is happening at ECHS and that teachers from both schools are talking together and helping each other rethink mathematics education at these schools.

CHANGING STUDENTS' ATTITUDES

The teachers at East Collins High School reported to us that a central feature of the PSML project is to change students' attitudes so that they are open to (a) accepting the challenges of mathematics, (b) seeing how mathematics is connected to the real world, and (c) taking more mathematics. We observed that teachers were attempting to bring about these attitude changes through group work and requiring that students take more responsibility for their learning.

Working in Groups

Being a member of a group is a typical student experience in the PSML project classes. As we observed classes of the 11 members of the mathematics department, group work was used in varying formats and for various lengths of time within the class period. Because of the consistent use of group work in mathematics classes throughout the day, the lack of group work by those classes reviewing for tests made them stand out as unusual. Group work was thus assumed to be part of the normal instructional day.

The teachers reported that the students liked working in groups and seemed more motivated to mathematics when working collaboratively. One teacher noted,

> Any time they [the students] can work together, they love that. And they want to know every day, 'Are we going to work together?'…So it gives them those things to look forward to in math, where if they were all in rows every day, they may come in there dreading it because there was not the interaction.

During interviews, students also spoke about group work. One female student who had recently transferred into a PSML geometry class noted that "we work in groups in our class, it's not 'you sit at your desk and you do individual work,' we work in groups every day. We have three people in our group, and so we all work together to get it done." A classmate chimed in,

> You always have someone there that can help you and, like, if you're foggy in one area and they're good in that area, they can help you. And if you're good and they're foggy, you can help them. There's always the chance that someone is going to know something that you don't know.

The first student agreed, adding that she valued the fact that "you can just be yourself and talk." "I learn better that way," she said, "in a more relaxed atmosphere."

Taking More Responsibility

We observed that the East Collins teachers emphasized students' taking responsibility for their own learning. Many of the teachers using a new textbook series expected students to read lessons in their textbook before coming to class, and often when students had questions during group work, the teachers encouraged and expected the students to use the textbook as a resource and to talk with other students before seeking help from the teacher. One teacher noted, "I think everything ... would fall into place if we could just convince them that they are capable and they can accomplish whatever they set out to accomplish with appropriate effort and work."

Some of the teachers noted that they felt students were beginning to take charge of their own learning. The department chair confirmed this feeling with a student survey he gave to his PSML classes.

> I had them do [the survey] anonymously. It amazed me that so many of them said the same thing. It was almost like some of them were saying, "Well, I know I would have done better if I'd have worked harder. I just didn't want to work harder, so that's why I got a D or an F." That's right. Not a one said this was too hard for me or if it had been easier I could have done it—nothing like that. They took the responsibility, which to me was a real big thing, the responsibility for learning. The first person that it goes to is the student; you're responsible for learning and then we'll help and do all of these other things. They basically accepted that—the successful ones and the not-successful ones. That, I thought, was victory; a small one, but victory.

Impact of the Teachers' Efforts

The teachers reported that their efforts toward changing students' attitudes seemed to be having some effect. They observed that students were less anxious and were taking more mathematics classes during high school; this was also noted by students and the associate principal. One teacher observed, "They feel more comfortable with mathematics and therefore they will pursue those high-level mathematics." A colleague agreed, adding,

> I see them lack the math anxiety. The math anxiety has almost completely disappeared—the "Uh, I just can't do math—" is gone. Or the "Uh, I'll just never be able to get this." That futility is not there. Frustration is still there. But there are avenues, and they know that there are avenues to take—more than one. It's not a do-or-die, now-or-never situation. They have seen that if they just keep working, keep plugging away, ask someone else, ask and talk. I see a change in the new students in that they are reading—they're reading mathematics. Their fear of the word problem—you know that always happens at the end of problem number 40. There's always two or three word problems that are tacked on as an afterthought. Word problems don't make them get ill any more, because there's so many of them. And we explore that in our everyday talk, so it's no longer something to abhor.

The goal of having students take more mathematics courses during high school took several different forms. Given the primary goal of PSML to have all students learn algebra, East Collins removed all general and business math courses from its curriculum. As a result, more Algebra I classes were added and students who would not have studied algebra under the former system are now in algebra classes. Additionally, the teachers proposed eliminating the honors courses and concentrating on challenging all students mathematically. The honors courses were eliminated after the second year of the project.

The teachers and associate principal reported that because of the change in student attitudes as a result of PSML, more students are continuing on to take Algebra II or geometry after completing Algebra I. The associate principal stated that the school "went from 67% of our Algebra I students going into Algebra II the year before we implemented [PSML] to 96% going into Algebra II the first year after PSML was implemented. We felt that was an amazing statistic."

One teacher commented on this situation:

> Something else I've noticed and talked about with some of the other teachers here— there are students in my Algebra II class who five, six, seven, eight years ago would never have made it this far in mathematics. They would never have made it this far, and yet they're equally competing and doing well. They're being successful.

Teachers also reported that girls seemed to be more engaged in PSML classes. One teacher mentioned that she had read an article that cited research that teachers pay more attention to boys. "They raise their hands all the time, they're usually the ones commenting." She noted that as part of the PSML project, the teachers were trying to "avoid those traps," and work equally with all students. She added that "it's kind of interesting how it's turned around. The girls are doing a little bit better than the boys are."

It appeared that the teachers at East Collins felt that their efforts to change students' attitudes had made some difference. They had evidence of more students continuing on in mathematics after Algebra I and saw students as less anxious and more responsible for their own learning. Most of the 22 students we interviewed seemed to support the teachers' feeling. Two students, in particular, exemplified the change in attitude. Both of these students were categorized by their teachers as average students, yet both decided to take two mathematics classes, geometry and Algebra II, during their junior years in order to get an academic diploma. Before taking an Algebra I course taught with the PSML philosophy, the students had not thought of seeking an academic diploma. However, both students stated that they now believed it was something they could do, and they wanted to take more mathematics.

DOING MATHEMATICS

For teachers and administrators at ECHS, changing students' attitudes about mathematics appeared to go hand-in-hand with engaging students in doing mathematics in the spirit of the *Standards*. We saw the *Standards* being interpreted by ECHS teachers to mean involving students in mathematics as reasoning through critical thinking, mathematics as communication, and mathematical connections.

Thinking Critically

We observed that many teachers used questioning techniques and interactive discussions in their classes to try to get students to think more critically about mathematics. One Algebra I teacher used an activity and questions to have students explore probability concepts. She asked all the students to look at their shoes. She then asked all students who were wearing white shoes to stand up. Students counted how many people had on white shoes and the teacher recorded the number on the chalkboard as $N(W) = 12$. She then asked all the students who were wearing black shoes to stand and recorded that number as $N(B) = 12$. The teacher asked students how they could find the number of students who had on white or black shoes. After some discussion, the students noted that they did not need to count everyone, but instead add $N(W)$ and $N(B)$.

T: If you have on jeans, raise your right arm. Twenty-four? Okay. Now suppose that I want to find the number of students who have white shoes or jeans. How many do you think that would be?

S: (Students give several different answers.)

T: All right, I hear some different answers. Can you make a conjecture about how many students you think have on white shoes or jeans? Thirty-six?

S: I think it's twenty-four.

T: Okay. What do you think? How many of you think it might be twelve? Four people think it might be twelve. Okay, how many of you think it might be thirty-six? Three. Twenty-four? One, two, three, four, five, six, seven, eight, nine, ten. No idea? Five people. What I want you to do is see if you can draw a diagram of these two sets.

The teacher continued on in this manner, asking questions and having students make conjectures and test them by drawing diagrams. By the end of the discussion, the students had developed an algorithm that enabled them to find the number of items in a set regardless of whether or not the sets were mutually exclusive.

Another teacher mentioned that he liked to have students explain their thinking processes. "Rather than solve a problem, explain how you would solve it; write the steps that you go through to solve it or give them open-ended questions. Give them four numbers or four objects or something and say, 'Which one doesn't fit?' and then, 'Justify your answer.'"

Students reported to us that they liked doing things that made them "use their minds." One student recalled a class involving a turtle race in which the students were given information and on the basis of that information had to form equations and predict a winner of the race and write about it in a radio broadcast format. Her classmate added, "Yeah, he [the teacher] likes us to be creative." Another added, "We will take math problems and turn them into stories. And I love doing that because I love writing."

Communicating About Mathematics

We observed the teachers at ECHS promoting and providing opportunities for mathematical communication. Because students often worked together, oral communication was expected. In a prealgebra class, students were working with the concept of area and perimeter. The teacher asked students to work on a task: "Make as many figures as possible with an area of four square units." The students worked cooperatively with each other, using grid paper, to complete the task. Discussion in the groups centered around whether figures they formed were equivalent to other figures but in different positions and whether they had found all the figures possible. Following a class discussion and group justifications of their solutions, the teacher had the students consider the question "Do figures with the same area have to have the same perimeter?"

In their groups, students made conjectures and tried to justify them by finding counterexamples or by proving it for all the cases. This task was followed by another one in which the students considered the question "Do figures with the same perimeters have to have the same areas?" Throughout classroom activities such as this one, we observed students engaged in oral mathematical communication as they struggled to understand concepts, to make and test conjectures, and then to convince their group members and the class of their solutions.

Another example of communication we observed occurred in the computer classroom, where it centered on geometrical ideas and technology.

S1: Look [addressed to the teacher]. I drew a nongon; it has twenty sides.

T: (Looks at monitor) What prompted you to call this a "nongon?"

S1: That's what the computer calls it.

S2: Well, an octagon has eight sides and…

S1: This is a nongon because it has more than twelve sides.

T: Where does it say what a nongon is?

S1: Here (points to a pull-down menu on the monitor). Well, it's an *n*-gon.

T: What do you think the *n* stands for?

S1: (Shrugs.)

S2: Is it like a variable?

T: In what way?

S2: Maybe it stands for the number of sides.

T: Right. So a polygon with twenty sides is called a…

S1: Twenty-gon.

Other examples of mathematical communication involved writing. Teachers reported to us that they sometimes had students write explanations of how they solved a problem or of the relationship among several concepts. At least one teacher had his students do outside written projects in which students researched a mathematical topic and reported on it. Another teacher had her students keep portfolios that were assessed every 6 weeks. The assessment procedure involved a self assessment in which students wrote a paragraph assessing their strengths and weaknesses and noting areas of improvement and areas that still needed improvement.

Connecting to Real-World Phenomena

Another way teachers reported that they try to engage students in doing mathematics is through mathematics experiments that allow students to have hands-on experience and make some real-world connections. One teacher described an activity he used.

> We did an experiment about a week and a half ago where you look through different size tubes at a meter stick on the wall to see how much of the meter stick you can see. Then you graph your distance from the wall versus how much of the stick you can see. It's a nice linear graph …. Well, we're going to do the same thing with a video camera and a television, where you point the video camera at the television,

but have the video going back into the TV. So you have a recursive loop. So you have the same problem—"How far away do I put the camera?" And you get some really nice graphics out of that. If you rotate the TV camera a little bit as the image goes through the loop, the images rotate on the TV screen.

Other activities were also used. One teacher described an activity another teacher did. "He built a cube—a cubic foot cube—and they were talking about volume or something and he lugged in gallons of water and poured in just to see if a cubic foot really held this many gallons of water." Teachers also set up laboratory-type experiments such as rolling a ball down a ramp to engage students in gathering information and developing a model to explore the relationship between ramp height and rolling distance (Winter & Carlson, 1993). Several of the teachers at ECHS had participated in summer institutes at Exeter Academy and gained ideas for experiments that allow students to use their mathematical knowledge to build models for real-world situations.

Another teacher had students do research projects on different mathematics topics or people involved in mathematics. One student noted,

Another thing that we did last year that I wanted to tell you about was that he [the teacher] gave us a subject dealing with math and we went and wrote a complete report on it. Like, it was a whole six-week theme …. I'm glad I did chaos … it was kind of interesting to see.

One teacher described how he liked to help students make connections between mathematics and the real world:

I just had them take any newspaper or magazine and find any article that had any math-related information in it, bring either the article or a copy of it, and write a one-paragraph summary of the article and turn that in, too …. So it's getting the kids to maybe read some things and maybe see some things that they never would have done before.

SHARED AUTHORITY

We observed that the reform in mathematics education that was underway at East Collins High School was perceived locally as being successful, in large part because of the way that authority and ownership of the project was shared through a number of two-way connections. At the core of all these connections were the mathematics teachers at East Collins (see Figure 5.1). The teachers have shared authority relationships with business representatives, school administrators, each other, and their students.

Teachers Sharing Authority With Business

State Power, the state power authority, and Southern Mills, a textile industry, contacted East Collins because they were interested in buying a computer-based algebra program for the school. As the associate principal explained, "It would benefit our students and benefit State Power in the long run. They were looking for a way to impact education and Southern Mills could use the program [for adult education] if it was in place in our school."

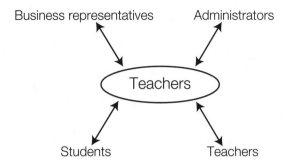

Figure 5.1. Shared authority relationships

Two mathematics teachers, the district mathematics coordinator, and the associate principal visited a school in the state that used the computer-based algebra program. The associate principal explained the teams' decision after visiting the school: "The school that we went to see was excited about the program; they felt like it was successful." The team discussed the computer-based algebra program but decided that "it was an expensive program; we would have gotten computers out of it, software out of it, but we just didn't think that it would fit our philosophy."

The associate principal mentioned that she thought that would be the end of it, because State Power had proposed only this specific curriculum package, but the State Power contacts asked, "Well, what can we do?" As one teacher commented, "We were just thrown completely off balance. Nobody ever says that—'You can do anything you want; What do you want?'" To answer this question, a team was formed.

This team developed a philosophy that formed the basis for the reform project. From reports we heard, it appeared that the teachers played a major role in formulating the philosophy and deciding how it would be implemented. State Power gave $55,000 for this project and the teachers had the authority to spend it as needed. This included buying textbooks that the teachers felt best fit the PSML philosophy, and graphing calculators. One teacher spoke of the empowerment gained by the teachers when they were able to design what they taught and how they taught: "We were able to use what we knew was correct and what we wanted to do."

Teachers Sharing Authority With Administrators

Teachers noted the support they received from administrators at the school and district levels in implementing PSML. An important endorsement came when the teachers requested that the mathematics sequence be changed for the third year of the project. That is, they wanted geometry to follow Algebra I, with Algebra II taking place in a third year. Prior to this, students took Algebra I, Algebra II, and then geometry. The district approved this idea, validating the teachers' authority over the curriculum.

Because of this change in sequence, there was an increased number of students taking geometry during the third year of the project; students who had taken Algebra I or Algebra II the previous year were now in geometry. However, this situation was only expected to last 1 year, while all the students who had already taken Algebra II completed geometry. The teachers reported that this led to some textbook difficulties; there was not enough money from the school district to purchase new textbooks for all the classes. In fact, only half of the Euclidean geometry classes had the new textbooks; whereas the other half had the textbooks that were used during the previous year.

> The chair reported the effect from this situation: A lot of the students who were in the classes using the old geometry books had been in the PSML program for two years, and they felt like they were being put out of the program. Our counselors were just swamped with kids from those classes wanting to be changed into the other classes .... it was a bit of a misconception there because the kids thought that the book was the program.

Teachers also expressed their frustration with not being able to use the textbooks that they thought most closely matched the PSML philosophy, and the associate principal went to the district office and was able to get enough new geometry books for all the classes at a time when textbook money was very tight around the state. The arrangement stipulated that the other high school in the county would be given the extra textbooks to use in coming years. The ability to have some authority over their classes was seen by the teachers as empowering. The fact that the district provided additional funds was a sign of endorsement of the efforts of the teachers through the PSML project.

The mathematics coordinator explained her willingness to share authority with the teachers:

> We believe the teacher is totally in charge, so you've got to have people in a position to know what to do and have confidence in those people to do the kind of job that needs to be done ... We give them leeway to develop and implement programs that they feel will be effective with their students. I think that's the way to go in education to be successful.

Teachers Sharing Authority With Each Other

During the first year, the project philosophy was tried out mainly by 2 teachers. More teachers came on board later, and by the third year all of the 11 teachers reported that they endorse and own the PSML philosophy. One teacher explained,

> I think one of the things that makes our math department so successful—I think we've got a great math department—is that PSML isn't something that began last year here. It's been in the minds of these people—all of us—in one way or another for years and then it's just that PSML is the focus of that.

We observed that the teachers felt ownership of the project and of the changes that have come about in their mathematics classrooms, and they seemed to value the contribution that others have made in this process. Several teachers discussed the fact that some of the mathematics faculty have worked outside the field of education and that this enabled them to bring real-world applications into the classroom. Others mentioned that they learned a great deal from one teacher who had some training in using cooperative groups. Continuing the model of teachers being experts, one teacher was selected to lead the development of the geometry course for PSML. The chair noted that he could "then teach the rest of us."

The teachers appeared to rely on each other for advice and ideas about teaching with the PSML philosophy. One teacher expressed an appreciation for chances to discuss problems and solutions with colleagues: "We're all about the same and we all have the same problems .... we all meet in the hallways for our little mini meetings." Other teachers commented on the collegiality that is built as they sit and talk while eating lunch in the department chair's classroom.

We observed that control and authority for PSML were being shared among the department teachers as they continued in this evolving process of defining the type of classrooms they want, and that through this process teachers were gaining confidence in themselves. One teacher noted,

> I always felt stifled, and I always felt like I had to ... and when I got carried away, you know, I always felt like I was being bad. And now it's like I have the blessing. I can swing from chandeliers, so to speak. That's the change. I feel better as a teacher; I feel more confident as a teacher.

Teachers Sharing Authority With Students

As noted earlier, an important part of the PSML philosophy is helping students to become more autonomous as learners. This seems to happen as teachers have students work in cooperative groups, ask questions instead of giving answers, and expect students to read lessons in their textbooks to learn new content. The mathematics faculty as a whole spoke about the value of students learning to work in groups. One teacher noted,

> I do a tremendous amount of group work. This is also what I perceive as one of the better benefits of this new philosophy ... I have worked in the outside world ... one of my worst problems that I had to deal with was people who could not work together. They didn't know how to work together. They [former students] had always sat in their same desk, with their notebook, and plugged away—totally and completely hand-held and guided. With PSML, the kids are not hand-held, they're not babied along—they're working together. Even though sometimes they don't get along at first, I keep the group together long enough so that they can work around their differences.

Another teacher mentioned that the students began to see that the teacher was not the only authority in the room: "When they get to a question, they'll ask somebody else in the class before they'll ask me." Teachers noted that as students relied less on the teacher and more on their peers or on themselves, their confidence grew. One teacher observed that her students were now "able to conquer math." She continued,

> This gives them a kind of confidence also, and I feel like any job that they pursue anywhere, they're going to have that confidence in the back of their mind and they can pick up any tool—and technological thing, anything out there—they're going to

be able to pick it up and learn it probably a little bit more quickly than someone who doesn't have that confidence ... so I feel that our kids might be a little bit more successful on the jobs they pursue because of this confidence.

Several of the teachers mentioned that sometimes students have a difficult time initially accepting new roles that give them more responsibility for their own learning. One teacher reported that a student told her, "You're not teaching us." The teacher continued, "He was unhappy because I wasn't teaching him in the same old way that he expected. And it's just a whole different way ... I told him to hang in there." Another teacher added,

> Some students were not used to the major techniques that are used with [PSML] ... They were used to having the teacher introduce the math technique, do some examples, work at their desk while the teacher walked around and helped them. [What we did in PSML] was just the reverse of that—they had to think on their own, reading first. And they ask, "Why am I reading?" They had to do it on their own. "But you haven't explained it to me yet"—and then having to think through some things without that step-by-step direction. That's very frustrating for students, until they get used to that. They fight it.

We observed that the process of sharing authority in the classroom was a learning experience for both teachers and students but that teachers seemed committed to the process and students appeared to take hold of the autonomy they gained. One teacher summed up the thoughts of several teachers when she noted that what was "hard at first was being able to sit back and let the kids do instead of me do—having a lot more learner focus than teacher focus. But once I got used to it, it was more comfortable. And I've seen [the students] become more independent in their learning and a lot more confident with what they think they can do in math."

The crucial element in all the shared authority relationships that exist through the PSML project appears to be the fact that the mathematics teachers at East Collins High School are at the center of all these relationships. The department chair summed it up when he noted, "that's why our faculty accepted it so readily. They knew that [the other lead teacher] and I had major input from the very beginning. As more of the control came to us, they saw that and it wasn't like some outside group coming in and saying, 'Here it is.'" We observed that the teachers were able to share authority with each other and with students in their classrooms because they shared authority with the business representatives and school administrators.

BRINGING IT TOGETHER

We observed that the teachers saw PSML fitting well into the vision of the *Standards.* One teacher noted,

> I see that PSML and the *Standards* are almost one and the same .... the *Standards* are what we are employing. It's PSML that gives us the routine by which to enforce the *Standards*, and the textbooks from UCSMP give us the hard copy. PSML is a philosophy; the hard copy is the textbook.

Another teacher mentioned that "the *Standards* really backed us up a whole lot" because she sees the philosophy of both documents as being similar.

> ...Parallel is a good word for it. In fact, I was doing a write-up for somebody and a question was "How do you use the *Standards* in your curriculum?" And when we first started PSML, we had to prove to the parents that [it] was good, so the *Standards* were right there. And we were showing them this: "Look, we are doing what's new, and what's out there, and what's good for your students." And so we use the *Standards.*

She felt that the textbooks they had chosen "fit the *Standards* so well" and promote a pedagogy wherein "we should no longer be a lecturer, or a giver of information; we should be a facilitator." As observers, we saw mathematics as reasoning and communication, and mathematical connections as prevalent ideas throughout the PSML classes.

We noticed that although each teacher in the mathematics department at ECHS had a different interpretation and implementation of PSML in his or her classroom, the underlying philosophy appeared to be shared by all the teachers. One teacher noted that PSML allowed for individual differences among teachers.

> ...PSML is very individualized. I think ... it depends on the teacher, and how that teacher is. [We] are different people, and the way we handle our students are different .... PSML is the idea that we bring into [the classroom].

It appears that PSML has provided the focus and framework for bringing about a number of things that "were in the works." The department chair noted:

> I feel like a lot of these things would have been done anyway—just from teachers getting together and discussing, and with the *Standards* and the graduation requirements changing in the state [to include a requirement that all students complete algebra] and the graduation test ... a combination of things, PSML included, contributed to some of the curriculum changes. PSML is about focusing people's ideas and giving them an outlet. The idea of evolving is very accurate. We talk about that among ourselves that every year it's a little different .... And I have no doubt that it's bigger than all of us. As long as there are a few to carry on, the others will pick it up as they go.

The idea of a program that evolves is evident both in the content changes and teachers' knowledge of better classroom practices. Some teachers at ECHS recognized a need to develop assessment practices aligned with the pedagogical changes that were taking place in the PSML program. They also realized that changing assessment from paper-and-pencil tests and grading only students' individual work to other forms of assessment would be a difficult and slow process. One teacher noted,

> Assessment is still something I'm working on. I've been worried about that all year. I kind of feel old, set in my ways, sometimes. But I've been doing a lot of reading lately and thinking about assessment, but I really haven't done too much about it yet.

Our observations indicate that teachers, administrators, and business partners who share in the philosophy of PSML through this shared authority expect PSML to continue to evolve and spread to other classrooms as teachers grapple with issues such as assessment.

Chapter 6

The Growing Pains of Change: A Case Study of a Third-Grade Teacher

Laura Coffin Koch

The introduction to the *Curriculum and Evaluation Standards for School Mathematics* (NCTM, 1989) lists five general goals that should be reached by all students. The *Standards* suggest that students who participate in a school mathematics program designed and implemented with these goals in mind will gain "mathematical power." Creating a classroom environment that sustains these goals and the spirit of the *Standards* is indeed a challenge. But as teachers go about making the changes, what happens to the school environment and teachers' relationships with those around them who may not support their changes nor share their beliefs about the teaching and learning of mathematics?

This is a case study of Barbara, a third-grade teacher who participated in a school district staff development program that dramatically changed her own teaching but at the same time generating tensions within her school community. As Barbara worked at her teaching and attempted to follow her beliefs to fulfill the requirements of the district program, she needed to deal with the conflict that arose as she changed. Teachers in her own school who were not experiencing similar changes seemed to value her less as a colleague. As Barbara struggled with the feeling of isolation, school and district personnel found it difficult to support her.

BACKGROUND AND HISTORY OF THE DEVELOPMENT PROGRAM

Five years ago, the Parker Springs Public Schools (PSSD), under the guidance of Angela Stevens, program specialist for elementary mathematics, received a large grant to develop a district-wide mathematics specialist program for teachers of Grades K–3. The intent of the program was to provide in-depth in-service and curriculum materials for one K–3 teacher from each of the elementary schools within the district. These teachers would then serve as agents for mathematical reform in their own schools.

During the first year of the project, 10 mathematics specialists participated in a pilot project. The next year, while the district waited to hear about the grant, no new specialists were chosen and one specialist moved out of the district. At the beginning of each of the next 3 years, 60 more teachers were selected as mathematics specialists. It was the goal of the district to have a mathematics specialist in all 90 elementary schools by 1995.

Although the initial grant proposal was written prior to the publication of the NCTM *Curriculum and Evaluation Standards*, the project attempted to align itself with the intent of the *Standards*. In 1991, the district published its own primary mathematics curriculum, modeled after the *Standards*, to serve as the school district's mathematics guide for all elementary school teachers. It provides the goals and objectives of the district mathematics program, as well as activities that teachers can use in their classrooms. Three of the 10 teachers who served on the Curriculum Committee were from the mathematics specialists group.

The program specialist for elementary mathematics and her assistant have provided ongoing support for the K–3 mathematics specialists by way of demonstration lessons and classroom observation. According to the program specialist and the mathematics specialists, strong social bonds have developed within the group. Each specialist keeps a journal that serves as a dialogue between the specialist and Angela Stevens or her assistant. The nine original mathematics specialists spoke about a "special" support system that has developed among them over the past 4 years. According to the specialists, they are there to "congratulate each other on their successes, to console each other on their perceived blunders, to share their ideas, and to collaborate" with one another.

TEACHER CHANGE AND TEACHER LEADERSHIP

The essence of the district's reform program is to rely on teacher change in the primary grades as the mechanism for bringing about change in the district's mathematics program. According to the program specialist for elementary mathematics, there are at least two arguments for relying on this strategy in the context of PSSD.

1. Although the district has developed a core curriculum guide based on the NCTM *Standards*, the district's site-based management program means that each school has the final say on curriculum.

2. The district administration has been impressed by the extent to which teacher change has been effected by a writing-process initiative in the primary grades and by the consequent effect on children's writing. The superintendent recalled listening to 5-year olds "critiquing each other and talking about style and form," sounding "like a group of college people sitting around." According to the superintendent, the values of the writing program transferred to the mathematics program:

> We just deal with expression and let them continue that creativity and worry about the mechanics later. That's been so successful that it laid some groundwork for change in the mathematics program the people needed. They learned there's more than one way of doing something.

Creating the environment that would enable teacher change was left to the program specialist for elementary mathematics. At the time the program was being developed, the district had embraced a three-part philosophy of teaching and learning.

1. There needed to be a shift from product to process and this called for teaching to be hands-on and child-centered.

2. Any new program had to be based on the most effective components of a staff development model.

3. Because the district already was using a site-based management model, the proposed mathematics program was to be centered on a school-based change model of change.

As the program evolved, the program specialist for elementary mathematics used the three criteria above as her guiding principles in designing a program that would—

1. Create a cadre of primary-level classroom teachers who were not only knowledgeable of the *Standards* but who also held the belief that children need to be the center of learning and that teachers need to be colearners in the mathematics classroom;

2. Use these teachers as mentors to new specialists and as agents of change within their own schools; and

3. Provide ongoing support to these teachers as they worked through the change process themselves and as they struggled with change within their own schools.

To this end, during the first year, the program specialist for elementary mathematics provided the K–3 mathematics specialists with over 100 hours of training that focused on problem solving, spatial sense, change theory, and the use of technology (calculators and computers). Well-known mathematics educators conducted workshops and seminars. The model adopted by the district to promote teacher change appeared to be well thought out but not so rigid that it precluded adaptation as the need arose.

The roles of the mathematics specialists are defined for each of the 3 years of the program. In year 1, the new mathematics specialists attend numerous workshops and seminars. Their focus of the first year is on themselves and their own teaching. In addition to attending the workshops and seminars, they meet with other mathematics specialists informally to talk about their own teaching and to share ideas and experiences. In April of the first year, the mathematics specialists attend the National Council of Teachers of Mathematics annual meeting.

During the second year of the program, the mathematics specialists continue to attend workshops and meet with other mathematics specialists, but are encouraged to invite other teachers in their schools to visit their classrooms to observe them teaching. The second year, specialists attend the state Council of Teachers of Mathematics annual meeting. They are encouraged not only to learn from these meetings but to present and share their own stories with others at their schools, in their districts, and at the regional and national levels, as well.

The third year of the mathematics specialist program is the time at which the mathematics specialists begin to share what they have learned in a more systematic way with the other teachers in their schools. This may be accomplished through formal workshops, seminars, short presentations at faculty meetings, or lunchtime discussion groups. Teachers are encouraged to visit the mathematics specialists' classrooms and to invite the mathematics specialists to visit their classrooms.

Throughout the program, the mathematics specialists are required to keep journals of their experiences in the program and in their own classrooms. Either Angela or the assistant program specialist reads and reacts to each of the journals.

Becoming a mathematics specialist was originally a 3-year commitment for teachers; however, most of the mathematics specialists continued working with the program in some capacity after the initial 3 years.

Although the grant covered the training and support of the mathematics specialists, the program specialist for elementary mathematics felt that it was important to provide additional training to Grade 4 and 5 teachers. By reallocating other support money within the district, Angela was able to expand the mathematics specialist program to create a Grades 4–5 liaison program. In this program, a K–3 mathematics specialist would be paired with a Grade 4 or Grade 5 teacher. The Grade 4 or 5 teacher would attend the same training sessions provided for the K–3 specialists. It was the intent of the program that the K–3 mathematics specialist and the Grades 4–5 liaisons would work closely together to help other teachers in their schools.

FIRST STEPS TOWARDS CHANGE

Barbara was a kindergarten teacher in 1989 when the K–3 mathematics specialist program began and was one of the first 10 teachers in PSSD to become a K–3 mathematics specialist. Barbara had been actively involved in the district's writing program, in which students were encouraged to write to express their ideas rather than focus on spelling, grammar, and punctuation. Barbara was eager to translate what she had learned from the writing project to her teaching of mathematics. As Barbara began attending workshops and seminars, she focused on compiling new mathematics activities that she could do with the children. Later, however, her focus shifted from amassing teaching activities to learning how she could teach mathematics better, how her students learn and her own learning of mathematics.

At the end of her second year as a mathematics specialist, Barbara reflected back on that year:

> Professionally, I have grown more this year than any other year in my teaching career, possibly more than all of those years combined! As a result of being exposed to so many experts in the field of mathematics, I have internalized the importance of making math meaningful for children. I have learned that there are no limits to one's ability to acquire knowledge and there's no set curriculum for any grade level. Allowing children to be producers of curriculum makes math become real and the children can discover far beyond my own agenda and expectations.
>
> This year, my class has been exposed to so many concepts that I never dreamed kindergartners were capable of understanding—multiplication, division, fractions, odd and even, counting in Spanish, counting by twos, fives, and tens along with addition, subtraction, complex patterning, measurement, time and money, ordinal numbers, one-to-one correspondence and counting by rote.

The children create math wherever they go—inside and out. From discovering a pattern as they jump up the stairs after recess to figuring out the daily lunch count and attendance. They even count and compare tater tots! One observer commented: "I couldn't believe your 5 and 6 year olds gave real numbers when they predicted how many olives were in the jar!"

In our classroom this year, we had 24 children as learners and one teacher-as-learner. Many times, I didn't know the answer to the problems that came up. We struggled together and learned that the answer wasn't the goal, but that the brainstorming that was involved to justify our solutions is what problem solving is all about.

I am confident that I am sending 24 "junior mathematicians" to first grade! It's going to break my heart when they walk out of room 24 on day 180, but I know they'll continue to stimulate new learning wherever they go.

Barbara shared what she was learning with the other kindergarten teachers at Armstrong Elementary School. By the end of the year, the kindergarten teachers were finding and developing their own activities and had abandoned textbooks. As a group, they began planning and sharing ideas with each other. By the end of Barbara's first year as a mathematics specialist, the other kindergarten teachers wanted their mathematics instruction to have more hands-on activities and problem solving and be less worksheet oriented. The kindergarten team decided not to order workbooks for the next academic year but to spend their money on a reformist textbook.

As we listened to the kindergarten teachers talk, it appeared that Barbara had made an impact not only on her colleagues' teaching but also on how they worked together and shared ideas. Barbara appeared to have passed the mathematics specialists model on to teachers in her own school. Two years later, these teachers continued to make efforts to work together and share ideas. The kindergarten teachers find time, at least once a month and/or during lunch, to get together to try out new ideas with each other and plan activities and lessons together.

MOVING ON

In 1992, Barbara moved to the third grade, a move that she felt was necessary to keep her vital and to avoid burning out. Barbara had taught third grade before coming to Armstrong Elementary School and wanted to implement what she had learned with older children.

The walls of Barbara's room were covered with mathematics posters and graphs; three-dimensional models hung from the ceiling; boxes of manipulatives filled the shelves; calculators were displayed in a hanging banner; and a computer and printer occupied the front table. On the board was a problem labeled "Problem of the Day" that read as follows:

We gave Fido a bath with a new flea-tick shampoo last night. We saw 28 fleas go down the drain. How many fewer legs were crawling on Fido after the bath?

Fido was Barbara's dog and she often shared stories and adventures about Fido with her class. Barbara talked of her penchant for finding mathematical episodes in her world and that of her students so she could bring mathematics to her students in a very real way.

Barbara: What do we need to find out?

S1:   How many legs on a flea?

Barbara: How can we find out?

S2:   It's an insect, so it has six legs.

Barbara: Now figure out the rest.

The students worked on the problem, then began to discuss how they arrived at their solutions.

S3:   Twenty-eight minus six.

Barbara: Does that make sense to you?

S3:   No.

Barbara did not tell the student he was wrong, but wanted the student to think further about whether subtraction would make sense in this problem.

Another student reported that he couldn't "do times without a calculator." Eventually, three students went to the overhead projector to show how they solved the problem.

The first student wrote "$6 + 6 + 6 + \ldots + 6 = 168$" on the overhead. The student added 28 sixes. She got the correct answer on her paper, but while doing the problem on the overhead she made an addition error early. She found and corrected her error and then stayed with the problem until she arrived at her original answer.

The second student went to the board. She said she knew how to write the problem, $6 \times 28 = 168$, but that she didn't know how to do multiplication, so she added: $28 + 28 + 28 + 28 + 28 + 28 = 168$.

The third student multiplied 6×20 and 6×8, added those results together, and got 168.

The students discussed their various strategies for solving the problem and the relationship between multiplication and addition.[1]

This type of problem was part of Barbara's opening routine each morning and not part of the mathematics lesson. In this third-grade class, multiplication had not yet been introduced.

As Barbara began changing how she taught mathematics, she started integrating mathematics with other content areas, especially reading. Barbara found books with mathematical themes, such as *Grandfather Tang's Story*, *Zero the Hero*, *The Doorbell Rang*, and *The Grouchy Ladybug,* and read them to her class. Barbara's students talked about how mathematics was more than problems on worksheets. "Math is everywhere," they discovered.

The second visit to PSSD took place 10 months after the first visit. When we returned to Barbara's room for this visit, we noticed our names written on the board, welcoming us and informing the students that we would be visiting the class that day. Barbara was working with the class on a variety of multiplication facts that on first inspection seemed to be low-level and unrelated. As Barbara

---

[1]This episode is included in Koch and Driscoll (1996). The mathematician Hyman Bass (AERA discussant) was compelled to write a mathematical commentary about the episode. (See Appendix B.)

continued with the problems, it became apparent that the students were giving her multiplication facts based on their names—the number of letters in a student's first name times the number of letters in a student's last name. As students gave their mathematics problems and the teacher wrote them on the board, other students checked them using calculators, pencils and paper, or mental calculations. When all the students had given a problem, we were introduced to the class. One child said aloud, "I wonder what's the value of their names?" Barbara replied, "What a good question! Let's figure it out." Barbara worked with the students as they figured out the value of our names.

Barbara talked of not being afraid to follow the path set by one of the children. The class had just started to look at multiplication facts by focusing on the product of their names. She believes that students' capacity for learning is far greater than most teachers believe. To Barbara, it wasn't enough to just glance over the question, or the solution; she sought to sustain the discussion and work the problem through until the students were satisfied. The class stayed with the problem, and it appeared that no one was left out of the discussion.

Barbara attempted to integrate mathematics into other content areas. She wanted the students to see mathematics as a useful tool in their lives. She did not want her students to believe, as she once did, that mathematics was cut-and-dried.

The following episode took place as the class was preparing for silent reading. When all the children were ready with their reading books, Barbara asked the students where the class had left off yesterday. After a short discussion, the class came to a consensus. The class then got into a discussion about how many pages they were going to read if they were to read one chapter. The discussion focused on how to include pages that were not completely filled with words. In the end, the students figured out that there were fractions of pages and then were able to add these fractions together.

Barbara's mathematical training was not that different from that of the other mathematics specialists. They all came into the program with varied backgrounds and prior experiences in their own learning of mathematics. Barbara felt that mathematics was not a strong point of her teaching and she had taken the minimum mathematics requirement in college. When asked her feelings and beliefs about mathematics, Barbara commented,

> I just thought it was an isolated subject that you taught from a textbook. In fact I thought when Parker Springs went middle-school concept, I thought I would like to teach middle school math. I wouldn't have any problems with the parents because math is either right or it's wrong, there's no in between. I had a totally different view of what mathematics was. I thought it was teaching algorithms, drill the algorithms, and move on to the next day. But now it's different. The kids are doing all the inventing of the algorithms and the inventing of mathematics. It's very different. It's almost backwards.

Although Barbara's previous experience in mathematics was limited, she is a risk taker, and not afraid that children will present her with a challenge she can't address. She believes that she is a learner and is willing to learn along with the children. Barbara talked of how much she believes the children can and do learn.

We talked about how she could assess students' learning, as she doesn't use tests or any other standard measures, and how she informs parents of children's progress. She described the portfolios her students are keeping:

Each of the students has a three-ring binder divided up by subject areas, and each subject has a table of contents. When I return work to them, they enter the date and the title in their table of contents in the subject area. Parents are always welcome to come in and see their three-ring binders. We have to negotiate where we put the material. There are a lot for subject areas. What is it? Is that reading or math? To me, that's a compliment that everything is that integrated. They put their work in and then at the end of the nine weeks, they do a lot of self evaluation. All they have to do is look at their table of contents to see. We staple two masterpieces that they have chosen to their table of contents for that nine weeks in each subject area and then put those in their portfolios. They also write newsletters home every week to their parents to let them know what's going on in the classroom. That helps and it has to be signed and returned on Monday.

At the end of the year, Barbara asked her class to write letters to her that would answer the question "How is math different in third grade?" The following examples are representative of the letters she received.

Dear Mrs. Myers,

You have made my life wonderful this year by changing the way that we do math by not giving us a math textbook. I really enjoyed that and I bet you enjoyed that too. The past years I've been using a math textbook and the problems were easy. Third grade is more challenging because we've been doing [Shaquille O'Neill] problems and measuring problems too. And class problems that deal with the kids in our class are my favorites.

Dear Mrs. Myers,

1. Math is everywhere.
2. Math is very cool and good.
3. Math is not boring.
4. Math is very exciting.
5. Math was not very exciting in the past years.
6. Math is good in third grade because there are no textbooks.
7. I love it very much.
8. Math is a lot of fun.
9. Math is in books we read.
10. Thank you for making our year in math good.

Dear Mrs. Myers,

In the third grade we have no textbooks and in the past we have had textbooks. In the textbook, you can't do football games. And we do lunch room graphs where we figure out who's absent and all of that stuff. We do # of the day and guestimation jar, and Mrs. Myers gets really involved with our work. It's more exciting than in the past years. We do mental math, lots of activities, and we work in groups. In the past we have done pages at home and this year we do something from a thing we have done in class. In the past it was the answer that was most important, not the thinking, and this year it's the thinking that is most important and not the answer.

This was Barbara's first year teaching third grade and her fourth year as a K–3 mathematics specialist at Armstrong Elementary School. The endorsements given by her students reflect the praise she receives from the school administration, from the parents of her students, from district personnel, and from other teachers in the district. Barbara, however, does not find this praise comforting. She only feels the tensions that are arising within her own school. It is those tensions and internal conflicts that are the essence of Barbara's story.

SPREADING THE WORD

Part of the mathematics specialist program calls for a link between the K–3 program and the 4–5 program within each school. At Armstrong Elementary School this translated into the "math buddies" program for Barbara and her fifth-grade liaison, Carin Lee. "Math buddies" evolved from a district program called "reading buddies." The program pairs a teacher in one grade with a teacher in another grade. The teachers meet to plan an activity that would be appropriate for both grades. One example that was observed involved third-grade and fifth-grade students working together on a tangram activity.

Combined groups of third- and fifth-grade students built tangram figures according to the descriptions in the book *Grandfather Tang's Story*, which features the construction of animal figures (bird, fish, etc.) with tangrams. Students were grouped in pairs (one third grader with one fifth grader) and the children worked as a team to construct various tangram figures. One teacher read the story and the other teacher displayed the figures on an overhead projector. At the end of the lesson, the students were to make a set of tangrams out of a piece of white bread and recreate one of the figures from the book. During the reading of the story, at least one pair of students made animal shapes different from the ones given on the overhead, but representative of the same animals. The teacher noted that her "kids did that a lot." She ascribed it to their work with the Lego Logo computer lab. In this lesson, it was difficult to discern which child was a third grader and which child was a fifth grader, as all children were equally involved.

Two years earlier, when Barbara and Carin worked with kindergartners and Grade five students, the older children helped the younger children with an activity. The teachers believe that the "math buddies" program benefits not only the children but also the teachers because it provides an opportunity for teachers and mathematics specialists to share and work closely with another teacher in the development of a lesson.

Carin talked of her limited interest in teaching mathematics, and said that in fact she did not really like mathematics but was willing to work with Barbara for a year. After observing Barbara and spending time talking with Barbara as they planned joint lessons for their combined classes, Carin began to see what the teaching of mathematics could be like. Carin started modeling her mathematics teaching after Barbara's. They valued their time together; not only was Barbara helping and teaching Carin, but Barbara was learning from Carin. In the past, Carin would have taught mathematics at the end of the day. That way, she would not have to teach mathematics if time ran out. That had all changed. At the time of our observation, Carin, like the mathematics specialists, made the most of mathematical opportunities that arose in the course of a day. Although the students in her class didn't use a textbook on a regular basis, they did use textbooks as resources. On one occasion, some of the students found an error in the teacher's edition. The following is Barbara's account of how her class handled finding the error.

> There was a wrong answer in the math book. The children felt we had to write a letter to them and give them the right answer. We explained why their answer was wrong and our answer was right, but the company never responded.

The kind of teaching that makes students excited about learning is contagious. Mathematics became an integral part of the entire school day in Carin's classroom, rather than a subject taught for 45 minutes or so. Barbara and Carin believe that mathematics should not be isolated to just one time period during the day but that teachers should take advantage of mathematical opportunities as they arise.

Barbara talked about Carin's growth as a mathematics teacher over the last 2 years and how she had observed Carin taking advantage of mathematical opportunities:

> When she first came to Armstrong Elementary School last year, it was very hard to plan. I had to say, "Come on, help me Carin, let's do something, what can we do?" She was a little inhibited by kindergartners, but now that I'm in third grade, she's more comfortable and we're able to do more. The focus just completely switches around. We had a man come in and read the Cajun style "Night Before Christmas," and we served doughnuts and hot chocolate. But they made a mistake when they gave us the Dunkin Munchkins. They said they only bought 60 and we had 58 kids in the classroom, but I knew when we opened up the box there were more than 60, but Carin said, "I only paid for 60." Well the kids ended up getting three doughnut holes each, and that ended up being this huge problem about how much she paid for them and how much each doughnut hole would cost. We wrote a letter to Dunkin Donuts, and the manager wrote back. But when we think we're just going to hear the Cajun style "Night Before Christmas," it ends up being so much neater than we had planned. But that's giving them the space to do that and not saying, "Well, I paid for 60, big deal. You each get two more, no big deal," which we could have done. But Carin is a thinker too, and she's really into saying to her children, " What do you think we should do?"

Barbara talked of helping Carin rethink her own mathematics teaching and the friendship they formed as a result of working and attending workshops together. Because Carin was attending the same types of workshops as Barbara had, they formed a bond originally based on teaching mathematics. Barbara told us that she found in Carin the support she needed and someone who shared her philosophy of teaching. At the same time, Barbara was beginning to sense that other teachers might not be as supportive of her efforts. As long as Carin was around to talk with and listen to, Barbara was able to "push aside" the feelings growing inside her about how she and her teaching were perceived at Armstrong. At the end of the year, Carin moved to another state to get married. When asked about Carin's move and any changes that had occurred, Barbara commented,

> Some things are probably the same this year as they were last year, except I probably just didn't notice them because Carin was such a buffer. I probably was oblivious to a

lot of what was happening because she was just there and we were there for each other and carried on. We both had interns in the spring, and we opened it up that we would do demonstration lessons for all the classrooms that wanted us to come. We went around and did demonstration lessons and had a ball. We had a great time for the people that let us in. There weren't a lot of people that let us. There was nobody on my grade level that let us come in. There was a fifth-grade teacher that let us come in on three consecutive days, which is really neat. We had a good time. She participated and seemed to really get a lot out of it. We went into another fifth-grade classroom, who's now in third grade, and we did a lesson on a chocolate bar. We did Marilyn Burns activities in her class and that was a lot of fun. I probably was oblivious to a lot of the resentment that I've noticed this year. When I'm here at school, I'm here at school for my classroom number one, and I don't sit in the lounge ... So I guess when I had an intern it freed me up to be with Carin, but I also began to see a little bit more ... but it really hit when I came back to school. And I never, ever anticipated it. I really, really didn't anticipate the ... the friction, you know?

When asked how this friction was expressed, Barbara responded,

You can tell by the body language. You can tell, I don't know—I'm overly friendly and I get excited. You can just tell by people's expressions. It's almost like they want to avoid me. It's hard.

Barbara talked of how she feels about comments made to parents by second-grade teachers regarding her teaching style and how that style would affect their children:

I never mind when parents come in. I never mind when other teachers come in. If they came in, they would see what was going on and not just say that she's not using textbooks and the kids aren't doing 80 problems with the same goal. But still it hurts when your colleagues say those things in the community, especially since not one second-grade teacher has ever walked through the door in my third-grade classroom. Not one of them. And all of them were here last year that expressed those feelings to the parents. In our school, parents are allowed to request teachers for their children, so you know the parents always ask their child's current teacher whose classroom they think the child would do best in. They couldn't get to everybody because I do have 11 children in my classroom this year that I taught in kindergarten, so that helps. And a lot of those children I taught in kindergarten have moved. I also have two children of brothers and sisters that I taught in addition to those 11. But still ... what do you say to parents that say, "My child's second-grade teacher recommended that I not put my child in your classroom because it's too over stimulating and you do hands-on?" And you know it just hurts, it just really, really hurts.

It appeared that the tension here was created by philosophical differences about teaching and how children learn. The second-grade teachers stated that students need to "know" the basics and that they need to be disciplined in their learning. To these teachers, this translates to students sitting quietly at their desks and working on skill-type mathematics problems. The difference in teaching styles was not only evident to the teachers (Barbara and the second-grade teachers) but also the parents (as expressed in the interviews) and the students (as noted in their letters to Barbara). Barbara does not tell other teachers how to teach, nor does she believe she has all the answers, but she does believe that these differences should not separate people, especially because all the teachers are working towards the same goal.

I truly believe that they must believe in what they're doing because we're all here for the same goal. We're all here for the good of children. So I don't know, but that was very, very, very, very hard.

ONE SCHOOL, DIFFERENT PHILOSOPHIES

During our first visit, the second-grade teachers talked about their concerns regarding the children's learning of the basic skills and consistency within the school's mathematics program. The following are characteristic of the comments made by the second-grade teachers.

> In second grade we have to start back doing basically things we used to do in first grade. Some children do not know their basic math until the second grade. If you want to do some of these problem-solving things, you have to know how to add and subtract to do those things, and I don't really attribute it to using or not using a textbook.
>
> I've been around for 20 years and taught in several different states. Twenty years ago I remember we were doing what Parker Springs is doing now and 5 years ago in Texas they were doing it. I've seen the same thing just roll around. I've seen a new idea come in and everybody jumps on the bandwagon instead of approaching it slowly and seeing how it works. Prescription learning for example, 20 years ago, is now coming back, focusing on the individual. We have a different name for it, but it's the same thing. Everybody jumps on the bandwagon and instead of doing a little bit at a time and seeing what works or melding it into the system with what else is good, we completely throw away what we had and do something new. That's what my concern is about. What we're doing now. In the classes I've taught, the children coming to me, I feel, are lower and lower and lower.

In our interviews with the second-grade teachers, we found that these teachers appeared to be traditional in their approach to teaching mathematics and believed that students need consistency and structure. Furthermore, they believed that the students were coming to them with lower and lower skills than ever before and that they really needed to work harder with these students on their skills. One teacher spoke of her concern for the students' learning of basic skills when innovation is brought into the classroom:

> Well, I don't mind speaking. I guess I'm a very traditional teacher. I've been teaching a long time and I've seen a lot of different programs come and go and I just find that in the last few years, I think that it's good to be innovative and to offer the children a lot of different aspects and different ways of doing things, but we're finding a lot of basic skills are really slipping away … being able to add and subtract, for beginners.

A second teacher talked about the continuity slipping away:

> Each person has his or her own idea of what direction to take math, but then the continuity is gone. One person believes that this is a certain direction to go, then there's a breakdown because in the previous year, those kids might have had a lot of experience with a different one. It seems the best experience overall was where the kids were put in a most balanced situation, where they had an opportunity to learn various methods of learning. Not just abstract and not just the manipulatives. They needed a balance between both and to be able to continue.

A third teacher discussed the effect of changes, particularly with textbooks, on the students.

This year we changed math textbooks. We were using Addison-Wesley and this year we changed to a new textbook and that threw the kids off. It's a whole different style of approach to math than the Addison-Wesley was. So when the kids were first faced with this new textbook, they were unsure. So I think next year's group, being more familiar with it, will find it easier to approach this. Even though we did the strategy and planning and the problem solving, it's emphasized earlier in the year in second grade, and as first graders they didn't see anything like that. So when they saw it in second grade, it really was difficult.

When the teachers were asked if they had a sense that the students were thrown off by changing emphasis and the text of the program, one teacher responded,

Whenever we do have different kinds of emphasis that the teachers, as well as the children, have a time of adjustment to see, as children go to different teachers and they have different methods of focus and direction then children have to get used to that, and I think sometimes, like she [another teacher] said, the continuity, the essential skills, the importance of what we're doing gets lost in the methodology.

These views provide evidence of the contrast between Barbara's ideas about mathematics teaching and those of her colleagues.

Barbara felt very uncomfortable with these differences. In her classroom, she continued to enchant, encourage, and develop "junior mathematicians." However, within the school, she began to distance herself from the other teachers. She asked not to work with another liaison after 2 years as a specialist, and declined to meet with Angela, the program specialist for elementary mathematics.

When Barbara was asked what she thought other teachers felt toward her, she replied,

I think the whole change process is so hard. In fact, I think a lot of it had to do with seeing all these people that come to my classroom, from all over the county. It's scary. ... you know, what are those people doing in there? What is she doing that I'm not doing? or Why isn't she doing what I'm doing? It was to the point where I would hear from parents. We live in this community and because of this, I know a lot of people in the community. I would hear from parents that they really wanted so and so to be in my class, but their teacher said they would just be so overly stimulated in my classroom because it's hands-on. I was safe to do that when I was in kindergarten because everybody thinks in kindergarten all you do is play, and that's OK. When I moved to third grade, all the teachers in second grade were highly, highly traditional. The second-grade teachers had been telling all the parents that this is where it starts in second grade. We begin with textbooks and it's drill and kill all day until your reading group is called over.

There seemed to be conflict between Barbara's role as perceived by the K–3 mathematics specialist program and that role as seen by the other teachers in the schools. The second- and third-grade teachers were unsure not only of the purpose of the mathematics specialists, but even about who they were. One teacher thought that they were a team of teachers just starting to come to the school. Another teacher believed that it would be hard to do what the mathematics specialists do without all the training and resources that are provided the mathematics specialists. There appeared to be concerns not only for all the material support provided Barbara and the other K–3 mathematics specialist but for the attention and recognition that Barbara received. During our first visit to Armstrong, Barbara casually mentioned that she perceived problems with her position as a mathematics specialist. During our second visit, Barbara discussed this with us in more depth:

> It was so hard that I even told my CRT (Curriculum Resource Teacher) when school started that I don't want any more observations in my classroom this year because last year it got to be so hard. One time I looked in the back of my classroom and there were 11 teachers observing me from other schools .... It's not that I mind people being in my classroom, but that's a distraction. Plus it's hard, and I'm not good at saying "no." But I didn't realize that the other teachers in the school seeing people coming in and out of my classroom was going to be a big inhibiting thing.

According to Angela and other mathematics specialists, Barbara had become well known in the district as an exemplary teacher. She wanted to share her ideas and to work with others who wanted to create a learning environment in which children would construct their own mathematics. Because of this, Barbara often had visitors in her room and was asked to participate in district-wide programs.

ADMINISTRATIVE INFLUENCES

The current principal and the assistant principal, Frances Smith, recognized and acknowledged the tension within the faculty as they considered the difficult issues of mathematics teaching and learning. The principal, although supportive of the program and of Barbara, gave the responsibility of working with the teachers and professional development to Frances, the assistant principal. Frances was very committed to the teachers' taking a leadership role in any change that occurs within the school and equally concerned that change not be directed from above. Frances told us she believes that there is a definite need to look at mathematics teaching practice in different ways, especially in the second grade, where she had heard more than one parent complain about the lessening of a child's interest in mathematics after kindergarten and first grade. She thought some teachers were holding back and needed to network and share ideas. She wanted this networking and sharing to occur naturally, believing the effects to be more powerful than if they were dictated.

Frances wanted to be able to provide a substitute in the classroom so that teachers would be able to observe and talk with Barbara and to visit each other's classrooms, but she also realized that the demands on teachers' time and energy are great. Just prior to our first visit, Barbara conducted an in-service workshop for the teachers. As a result of this workshop, a monthly seminar was organized on the basis of one of Marilyn Burns' books. Twenty-four teachers participated in the seminars that took place between our visits. Barbara was delighted that the teachers took this initiative and wanted to let the teachers know that she was committed to supporting and encouraging them, so she agreed to buy copies of the book for those willing to participate in the seminar. During our second visit, Frances reflected on the effect of the seminar:

I believe the last time you were here, we were just getting ready to start our math study group. That was exciting to see. I enjoyed that. I was able to sit in on a few of their sessions, not nearly as many as I had wished. I think it was really very beneficial. It was interesting to even watch the different dynamics of the groups. I think the primary group just immediately bonded together, really came together as a group and was very, very supportive. I didn't see quite as much of that in the intermediate grades. I think the intermediate teachers are not quite as ready, but yet they did participate. Even within the journals you could see that .... I had asked them as part of the staff development component to respond in the journal and some of them truly did regret that they just made a list of the activities that they did ... you could just tell there wasn't as much thought put into it. So that's been interesting to note. I've been real excited to see changes in our second-grade level. There wasn't as much happening and we moved one of our first-grade teachers there. At the end of the year, after the study group, I told them that if they had missed a session, they could view the Marilyn Burns tapes and get some additional points. That was really good because many of them who did said, "Oh, now I see it." After having read the book and seeing the videotapes, it put it into real life for them. One teacher came to me and said, "I really want to do this next year, but I'm going to need your help."

Frances felt that change was making its way with the second-grade teachers, and at the same time she was cognizant of the frustrations that Barbara was experiencing that year. She acknowledged the complexity of the relationships among the teachers and parents:

The professional jealousies with teachers. I don't know what we can do to eliminate that. And there is some of that. But I think it's also compounded by some other issues as well .... Parents, especially in this type of school, are a very, very big factor. And I don't know if Barbara shared with you some of her concerns with parents or not. I think she had felt she had lost some parent support from some people. However, there were just as many people, if not more, who requested her as a teacher this year, because their child had experienced her before and [they] were very pleased and happy with her. So the only problems I think we have with parents are those who truly don't know and understand what's going on. For those who do, there's no problem at all. They're very excited and very supportive. We've not gotten into any kind of big problems.

Although Frances did not see Barbara's unhappiness and isolation as a "big problem," to Barbara, it was career-threatening. Teaching is Barbara's love and her lifelong work. Frances was not comfortable with the tension, and she believed that headway was being made by the second-grade teachers. She did appear to not want to "rock the boat" by interfering. It may be that Frances did not know the depth of Barbara's unhappiness, rather than that she did not want to face it directly; or, she recognized the complexity and challenge of this type of change.

Frances appeared optimistic that changes were taking place that would help resolve the situation but recognized that those changes might come too late for Barbara to stay at Armstrong. In our last interview with Frances, she talked about the changes being made in the second grade, particularly with one of the most traditional teachers:

What's so interesting is Zita, a teacher on the second-grade level, went ahead and ordered math books this year, but I think she is probably further along than the others.

But I guess it was an additional security. It's been neat to see. I just walked in a classroom the other day of a teacher and they were doing one of the games with Mayan digits. She asked me to come and see what she was doing with the math menus and how her children were doing. I noticed that same game in this other teacher's classroom, and I commented to her that was great. I love that game. There's so much thinking that they have to do. She said: "I really feel guilty because we're not really doing math because they're having so much fun." But she was able to see. She said you can see all the strategies that they have to use and do and everything. I asked her how she got the idea and she said that she had gotten it from Zita and that Zita was starting to share things with her and that she'd borrowed the book because she didn't take part in the study group last year. It's just really neat to see it starting to filter on down and through and around. I'd say what is exciting is that they're still moving. They talked a lot about wanting to do another group this year.

Frances wanted the teachers to know that they are all valued, regardless of their philosophical beliefs. Through this process of change, she was trying to create a "community of learners" within the faculty and a sense that it is okay to disagree.

I've really tried to work very hard and I've not done as good a job with this as I should have. I try to let everyone know that I value them. I need to be supportive of everybody where they're at. However, a couple of people have made it very clear to me that they don't feel that I give them as much because of their instructional beliefs and strategies. We've gotten into discussions about things which we don't agree on. The one thing that I would love to be able to see is if we could just agree to disagree. That that's okay. It truly is. I don't have all the answers, nobody has all the answers. I think it all goes into building that whole "community of learners" spirit within the teachers in that community and a sense that none of us know the answers. The only thing that we can do is just to continue … There was just no way we were going to come to a middle ground on this. They got very frustrated because I wouldn't go that way. I'd tell them, "I'm sorry. I'm going to give you the freedom. But if you feel that's what's best, then that's what's going to work best for you."

THE IMPACT OF OTHER TEACHERS

Teachers in other grades expressed gratitude for the help and support given by Barbara. Barbara had worked with a first-grade teacher in the development of a classroom store and shared materials with her. Gina, a fourth-grade teacher who had many of Barbara's students from the year before, knows that Barbara is doing wonders with the students. Gina developed an activity for her class. The students were given a series of calculations related to the buying and selling of baseball cards and asked to write a letter to someone to explain an error in calculating the cost of baseball cards. She was convinced that activities like this provide "proof" that Barbara's students are better than those who did not have Barbara.

On our last visit to Armstrong, we again talked with teachers, including the second-grade teachers. It appeared to us that there had been a definite shift in their thinking and in the way they talked about teaching. They talked about using manipulatives, teaching children to think and to understand why, asking children for alternative ways to solve problems, and not focusing only on the answer.

Most talked of their attempts to use portfolios as a way to assess the students. They also discussed how they were beginning to feel more confident in being able to vary from the rigidity that a textbook-driven curriculum had placed on them. The comments made by a permanent substitute (one who stays in a single school on a regular basis) are characteristic of the comments made by the teachers:

> I don't feel behind because I believe it's what's good for them. I think it's much easier to see each child as an individual and not just teach to them as a group. Now I can go back and look and see which child has which weaknesses and therefore hold small groups in the summertime or something to work with them on their needs and before I may not know who couldn't do certain things. I feel I know my children much better doing the observation and get to their specific needs.
>
> Going into the different grade levels like I do, I found out, I sort of have to, like, read over real quick, and I come in early just to find out what the teacher has planned. I try to go by their lesson plans. If I'm trying to teach something and they don't understand it, I ask them, well what is it … what part aren't you getting? You need to make them think and to be responsible for being able to tell you, "I just don't understand all of it," or "I just don't understand how come if you multiply two times three, it equals six and then you add this. I don't understand where you're getting the six from." I'm putting the responsibility back on the child to find out what part am I not explaining for you to understand and then a little light comes on, Oh, okay, because then they explained it to themselves, really. When I'm making them break it down.

When the teachers were asked if they felt free to deviate from the textbook, one teacher responded,

> Yes, I really do. If there's something neat, I want to try. If I have some idea, I do it. I don't use the textbook. I should say I'm in a new grade level, so I refer to the textbook to make sure that I am covering the material that is supposed to be covered. That's it. But I do it my way. I try to read up and get as much knowledge as I can and go for it the way that I want to teach it.

One teacher talked of how change in her teaching began with the encouragement of the administration and other teachers:

> I've taught 21 years, so I know real different. And I think the reason I changed is because I've read things and we've been encouraged to do things by the administration and by other teachers. And, of course, I think it helps when other teachers encourage you and are willing to share and then you just try it. Even though I felt at the beginning of the year it wasn't real successful, I kept thinking they just weren't used to small groups, and they weren't used to partners and I kept trying. I thought, "I'm not going back to the old way of doing this. I'm going to stay this way." You just have to think and keep thinking that it will get easier, and it has. Now they're doing real well at it. But it was real hard at first. I wanted to change and I think that makes the difference. If you want to change and you want to try something, then you can.

We asked the teachers to reflect on what they felt would be important for their children to take with them with regard to mathematics. The following are comments made by several of the teachers.

> I would like for them to be able to tell you why they do math and show why they've done that. Not just to be able to throw it out, but be actually able to verbalize to say to you, "This is why I did it."

Their love of math. Just that math can be fun and [they can] feel good about themselves with math because they have so many more years to go with math and it's a lifetime thing. It's not just a subject, it's all around you.

I think they relate the term *math* with the enjoyment that they have in the classroom when they do ... and I make sure I call it math when we're doing it to so that when they get into the upper grades they don't panic when they hear the word *math* as something difficult that only upper-grade people do. They already know that they're confident in math and they've loved it all their life.

CHANGES IN THE STUDENTS

Changes in the mathematics program at Armstrong Elementary School had not eluded the students. Students in Carin's fifth grade talked about what math had been like the previous year and what it was like the year we revisited: "We opened our books and we did math pages." "It was pretty boring." "You forget it easier when you are doing it in the book." "Yeah, if you act it out, you can remember." "You can remember, because it's more fun. You can remember the fun things." One girl decided that she was not going to go on and if they told her she was she would say, "No. I think I'd rather go back to fifth grade now." Similar sentiments were offered by the students in Barbara's third-grade class when she asked the students to write about the differences between third-grade math and their previous math classes. "In the past years there was just one time during the day when we did math, but in third grade we do math all the time," wrote one. "Last year, math was 'Open to pages 50 to 53, do 1 through 10 and save the rest for tomorrow'; the problems were simple," another student recalled. Wrote one student enthusiastically, "In Mrs. Myers's third-grade class, the problems are challenging. I love it." Another observed, "In third grade, the answer isn't as important as how you get it. Sometimes there is more than one answer and my mental math is better. It is really challenging in third grade." Many students mentioned relief at not using a textbook, and almost all students commented on enjoying mathematics or finding it exciting and challenging.

We had an opportunity to interview both Barbara's current students and her students from last year. The interview focused on their notions of mathematics, how useful mathematics was to them, and issues related to their experiences in learning mathematics. We asked the students, "What types of questions does your teacher ask?" Barbara's former students responded that their teacher didn't ask many questions in math. Barbara's current students told us how Barbara always made questions that were "real-life":

> Remember when Fido took a bath, how many fleas came out and about there are six legs on fleas and then she asked us and there were four spiders in her house when she was sweeping the walls in her new house and then she checked and then she asked us how many legs have been crawling in her house?

We asked the children if their teacher (Barbara) could have them leave her classroom with one thought about mathematics, what they think she would want them to take with them. A third-grade student responded, "multiplication." This student had only been in Barbara's class for 2 months. Responses from the fourth graders (Barbara's students from the previous year) included comments relating to the idea that "math is everywhere." The following are representative of the children's responses:

I thought that she wanted us to know that math is very important because you see it everywhere you go.

I think she wanted us to realize that math is everywhere. Because in second grade, you weren't taught that and she wants you to realize that it's a natural part of life. There's a lot of things you couldn't do without math.

Barbara's students are not only learning mathematics, but valuing mathematics and beginning to see that mathematics is more than pages of textbook problems.

In our classroom observations of Barbara, we saw many open-ended questions asked of the students; emphasis on students' talking and writing mathematics; students working in pairs and small groups; a teacher attempting to understand the thinking behind the students' answers; comments to get the students to articulate their thinking; and the use of manipulatives on a regular basis. Students seemed to be engaged in the learning of mathematics.

Barbara talked about her reality. She knew that she had made great strides in her own teaching. It was evident in listening to her students, listening to the fourth-grade teachers talk about her former students, listening to the associate principal talk about her teaching, and hearing the requests from parents that their children be placed in her class. Unfortunately, Barbara also realized that she had been given a responsibility to promote change in her school and felt she had fallen short. Barbara talked with us about her feelings and her work as a facilitator of change:

People do still come and ask for things and, you know, and I always try to fulfill whatever they want. I don't want to be isolated, but the way I've begun to feel is like, She thinks that she's the only one doing it right. And I don't want people to think that. Therefore, I'm not going to people, they're more coming to me, where in the past I was just, like, flitting and buzzing around and very naive and just thinking the whole world was excited about doing and making change. Everywhere I went I was very excited. Now, I guess my self-esteem has really dropped this year with other teachers. Especially here. It's just like I don't want to push myself on anybody and I don't want anybody to think that I know all the answers because I don't. And when I do, that's when I'm going to quit teaching.

Clearly, the challenge of taking risks and providing leadership has been difficult for Barbara.

SUMMARY

In the many examples we saw of Barbara's teaching of mathematics, whether it was during her opening, at "math time," or at other times throughout the day, Barbara demonstrated her interpretation of the *Standards* and how the children affect the curriculum.

• Children were not only problem solvers, but problem creators. Each day,

Barbara presented the class with at least one problem to solve. In an example given above, a student questioned the numerical value of a name. Barbara encouraged the students to raise questions and then worked with the class to follow a question through.

• Mathematics is connected to other content areas and to the real world of both the teacher and the students. During our first visit to Barbara's class, she had been reading one of the "Ramona" books in which Ramona's mother decides she needs to buy tongue rather than meat because it will be cheaper. Barbara decided she would go to the supermarket to buy some tongue for the class. The butcher gave her the tongue without charge. While explaining to the class what she had done, she immediately made it into a mathematics problem by asking the children to figure out how much money the store had lost by donating the tongue to the class (4.28 lb. at $5.95/lb.). She did not hesitate to give her class this problem just because they hadn't gotten to multiplication or decimals yet.

• Students are thinkers and reasoners. Barbara continually asked the students what they were thinking, how they arrived at solutions, if there was another way to solve a particular problem, and if a certain solution was reasonable. Barbara used an activity that she calls "Number of the Day," in which someone selects a number and the children try to guess the number. The person selecting the number can only say whether the number is higher or lower than the number guessed. Once the class finds the number of the day, students give as many number sentences as possible that use that number as an answer. Students are to do these without a calculator, but may use a calculator to check or verify the equations given by others. All students appeared to be involved in this activity, many giving multiple equations. On one of the days that we were observing, the number was 333. Representative of student-generated equations were the following:

$$100 \times 3 + 33 = 333$$
$$999\,000 \div 3\,000 = 333$$
$$10 \times 10 + 200 + 33 = 333$$
$$1066 - 733 = 333$$
$$800 - 467 = 333$$

As Barbara wrote these equations on the overhead, students raised and waved their hands trying to get recognized to give their equations. Several students told Barbara that there were patterns that could be continued ($633 - 300, 733 - 400$, etc.) and others noticed that some were just shortcuts of others ($50 + 50 + 50 + 50 + 50 + 50 + 33$ could be written as $50 \times 6 + 33$).

• Discourse was the norm in Barbara's class. Students talked in groups and participated in whole-class discussions. In talking with Barbara's students, those who were current and those from a previous year, they stated that they believed that mathematics is powerful and that they possessed that power. The students in Barbara's class made pentomino shapes, according to a set of rules set up between Barbara and the class, and had to decide how many constituted a full set. Students discussed the meaning (and spelling) of pentominoes. This started with a discussion of dominoes and ended up with a discussion of pentagons. This activity was related to a previous activity in which students were to figure out all the possible combinations of squares that could be put together to make a closed box. Initially, students struggled as they cut the pentomino shapes out of inch graph paper. Many students confused the rules for making pentominoes with the rules for making closed boxes. Although the teacher told the class how many shapes constituted a set, the class was not ready to accept her "slip." They knew that she could be trying to trick them, and besides, they wanted to figure it out for themselves.

• Not only can children construct their own knowledge of mathematics, but teachers are colearners with their students. This notion is extremely powerful, especially as it relates to the teaching of kindergarten. Barbara exhibited a sense of excitement and awe as she listened to children explain their solutions. We were continually amazed at the power she granted the children in their learning of mathematics.

Children do not learn in isolation of their own experiences (Kouba, 1989). Teachers learn about children by interacting with the children, listening to them, and asking them questions. Understanding children's thinking is an important aspect of the mathematics specialist training program in which Barbara participated. In this classroom, Barbara and her students are colearners.

It appeared that, although Barbara could not see that she had made a difference, the school was moving in the direction of the *Standards*. These teachers conveyed impressive and presumably sincere passion about their work in mathematics education. The growth in just 9 months' time appeared to us to be substantial. Although most of the teachers were not as far along as Barbara in the process of change and some were just beginning, change was definitely occurring from within the ranks of the teachers.

The relationship between the program specialist for elementary mathematics, Angela Stevens, and Barbara Myers was a complicated one. Angela knew that Barbara was in the midst of an inner struggle at Armstrong Elementary School and was aware that Barbara was unhappy the year we visited, but was unaware of the cause and extent of her concerns. Barbara and Angela seemed unable to communicate their frustrations with each other. The associate principal was also concerned, but felt that these problems would work out over time. Interpersonal tensions add a layer of complexity to the already difficult process of making change in mathematics education.

There is limited mathematics education research about the psychological and sociocultural mechanisms of teacher change (Goldsmith & Schifter, 1993). However, we do know that change takes time. Teachers such as Barbara may need to be patient as other teachers ponder and effect change in their own ways, and as administrators decide how and when to intervene in the process. Barbara had been undergoing change for a while, first with the writing program and then as a mathematics specialist. Others will need more time to reflect on their think-ing and practice, as Barbara had time to reflect. Deep change must come from within a person and not from the outside. Despite Barbara's concerns, there were signs and inclinations at Armstrong Elementary School that change was about to happen.

ISSUES AND CONCERNS RELATED TO TEACHER CHANGE

Although the K–3 mathematics specialist program did anticipate problems for the teachers, could it actually prepare the specialists for those changes, especial-ly changes that related to socialization in the teachers' own schools? The chal-lenge, for teachers, of being agents of change within their own schools has been substantial in a number of cases. This model is, however, the basis of the PSSD K–3 mathematics specialist program. There are other examples of this type of model working effectively (Moon & Hein, 1995), so understanding its nuances and complexity is important.

Parker Springs' reform plan called for commitment from not only the mathemat-ics specialist, but also from the principal of the school. Changes in administration at Armstrong, coupled with Barbara's disinclination to "brag" or promote her own work, may have contributed to this situation. Mathematics specialists at other schools did not experience similar problems, especially when they had principals who were supportive of the mathematics specialist program.

The data collected in Parker Springs School District, particularly at Armstrong Elementary School, indicate that reform programs can take a great toll on indi-vidual teachers and their relationships with other teachers within their schools. Programs that rely on teachers as agents of change need to help prepare the teachers for shifts in roles and perceptions that may influence the day-to-day interactions of teachers within a school. Extra resources, attention, and acclaim directed to teacher leaders can create difficult dynamics among colleagues, as we saw in this case. Barbara was sensitive to how she had benefited as a teacher from the program and realized that her colleagues had not had the same advan-tages. Better communication strategies might have helped this tension.

If reform is going to succeed, it cannot be "top down," but needs to have a sense of shared ownership. In Parker Springs, the mathematics specialists did share in the ownership, as did some of the principals. This was the case in many of the schools, but this ownership came about over a period of time. Teachers understood the role of the mathematics specialist and the administration sup-ported the role of teacher as agent of change. The mathematics specialist was a resource person for the school, the teachers and the staff and was there to work with teachers who believed they could do a better job of teaching mathematics. District and school administrators can help mediate or prevent such tensions among the teachers by working with all teachers early on in a program and by providing training and support to help the mathematics specialists understand the role of change agent. Furthermore, school administrators need to advocate for the change agent and realize that feelings and perceptions are real and need to be acknowledged and dealt with. The mathematics specialists had different experiences and they all went through some type of trial, but at different times. District-wide reform projects would do well to have ongoing support for these agents of change to help them work through their individual struggles, whether those struggles have to do with acceptance by other teachers, standardized testing, or parental pressure. Programs that create teacher change and teachers as agents of change may be well-intentioned and bring out true teacher change, but the process is demanding. In Parker Springs, a number of schools were successful in bringing about widespread change. At Armstrong Elementary School this was especially difficult. Enacting *Standards*-based reform can include unexpected challenges.

Chapter 7

Highlights and Implications

Joan Ferrini-Mundy and *Loren Johnson*

The process of change in mathematics teaching and learning—for an individual teacher, a school, or a district—is slow and arduous. The visions articulated in the NCTM *Standards* documents are subject to multiple interpretations. The degree of "*Standards*-likeness" of a classroom is not amenable to convenient quantitative characterizations. By electing an ethnographic approach in the R^3M project, many of the difficult issues relating to monitoring reform have become explicit. In developing the mechanisms for studying 17 sites, with a team of 22 researchers of diverse experiences and viewpoints, we were faced with challenging decisions about what perspectives might guide our work. The overview provided in Chapter 2 by Schram and Mills provides an account of this decision-making process.

We agreed that we needed to use the *Standards* documents as a philosophical and conceptual base. In the same way that the *Standards* value the diversity of student thinking, this study needed to value the diversity of site interpretations. The researchers decided that in project sites they would work to understand the way communities, schools, classes, and teachers were choosing to interpret the directions and ideas of the *Standards* and do their best to tell the story so that readers might benefit from the combination of contextual background, description of the innovations, and description of the nature of the process. The case studies in Chapters 3–6 are the result.

The highlights that follow draw on both the case studies and the broader set of scenarios. To conclude, we discuss the implications of this study.

HIGHLIGHTS

As we pointed out in Chapter 1, there were five themes that guided the collection of data in the R^3M project:

- the mathematical vision held by the people in the site;
- the pedagogical vision, with regard to mathematics, articulated in the site;
- the contextual influence on teachers' efforts to change their mathematics practice;
- the influence of mathematical and pedagogical reform practices on students at the sites; and
- the evolution of mathematics program reform at the sites.

Using these themes as an organizational structure, and drawing on the scenarios, the case studies, and the cross-site analyses that have been undertaken by Johnson (1995), we provide highlights of R^3M that are interesting, noteworthy, or compelling in some way.

MATHEMATICAL VISION

Identifying and describing the "mathematical vision" in the research sites was the most difficult part of the process for the R^3M documenters, judging from their write-ups. They struggled to determine how mathematics content had changed in some systematic way within the sites to reflect the recommendations of the NCTM *Curriculum and Evaluation Standards for School Mathematics* (NCTM, 1989). Graham and Ferrini-Mundy (1996) identified five features of the mathematical visions that R^3M researchers found in the sites:

1. an emphasis on the processes described in the first four Standards: mathematics as problem solving, mathematics as communication, mathematics as reasoning, and mathematical connections (NCTM, 1989);
2. a view that mathematics is more than computation;
3. an uncertainty about how much emphasis should be placed on learning basic skills;
4. a great diversity in the expression of the vision by various stakeholders; and
5. a philosophy of "mathematics for all."

Individual teachers clearly held a variety of views about the nature and practice of mathematics and what content areas of mathematics are important for students to learn. For example, Dossey claimed (1992) that whether the view of mathematics as a discipline is "static" or "dynamic" greatly influences what teachers value as mathematics and presumably what they will pursue in their reform of their mathematics teaching practice. The R^3M study revealed that the varying perceptions of what teachers and other stakeholders held to be important in mathematics shaped the reform efforts and the mathematical vision of individual teachers and entire schools. Here we highlight one important tension that seemed to relate more to the practice and nature of mathematics than to the content of mathematics.

Skills Versus Concepts Tension

Some of the teachers we met expressed concern that "hands-on," exploratory activities for learning mathematics, with emphasis on conceptual development rather than skills practice, would circumvent the learning of the computational and algorithmic procedures that have dominated and continue to dominate the mathematics curriculum in Grades K–12 (Weiss, 1992; 1994). Generally their ambivalence about this issue seemed to stem more from genuine concern about the long-term well-being of their students rather than from strong views about the nature of mathematics and its practice. Examples from the case studies, especially Parker Springs, illustrate the tension the teachers felt.

Teachers' beliefs influence actual curriculum as practiced in the mathematics classroom (Robitaille et al., 1993). Some teachers felt that students must be grounded in basic mathematical skills before they could be allowed to explore mathematics problems. Other teachers believed that explorations create an excitement and interest in mathematics that lead to the need to develop specific skills. At one school, the department chair expressed confidence that the basic skills that are normally measured by standardized tests would be acquired as a consequence of the pedagogical shift to concept development and exploration. "I think we're going to see our scores go up, and I think it's going to make a lot of people very happy. I wish [testing] was not an issue." Another teacher said, "My personal opinion is that [our program] will make their scores go up."

On the other hand, teachers at an urban high school were concerned that administrators' words of encouragement for mathematics reform efforts might not necessarily translate into support, because teacher evaluations in the past were often influenced by how well students did on standardized testing. Teachers were worried that poor evaluations could result if they wandered too far from conventional practice, particularly if student scores on standardized tests fell during the early stages of program implementation.

Data from the study indicated that a greater emphasis on mastery of basic mathematical skills tended to occur in those sites that either had traditionally low or high scores on standardized testing, especially at the high school level. In one high school in the study, for example, student scores on ACT, SAT, and Advanced Placement exams were traditionally the highest in their state. There was a real concern among those at this site that reform that lacked a strong focus on skills development would jeopardize these test results. The teachers at another high school had the same concern about skills scores dropping, but in this case the reason was a concern about not having low scores fall further (Johnson, 1995).

It is not at all clear from the project in what way the actual text of the *Standards* documents either clarifies or contributes to the tension teachers feel with regard to this issue. Certainly the tendency seems to be to read the *Standards* (NCTM, 1989) as recommending a deemphasis on basic skills, although the document clearly argues for basic skills development in several places (e.g., p. 20) and only recommends that there be "decreased emphasis" on "rote memorization of rules" (p. 21).

Standards Topics as Add-Ons Versus Infusing the Standards

A very interesting question about *Standards* interpretation centers on whether practitioners view the *Standards* as prescribing a comprehensive outline for K–12 mathematics or more as a menu from which favorite topics and approaches can be selected. We found evidence in the R³M study of both interpretations. Examples of those differences were demonstrated in the cases of two of the high schools in the study.

At Desert View High School (DVHS), topics (projects) were appended onto an existing and full mathematics curriculum. The reader will recall from the case study that the teachers were intrigued by the use of projects in classes at a nearby university. After developing their own projects in collaboration with the university faculty, the projects became add-ons to the existing mathematics curriculum, with teachers assigning a minimum of two projects per year to their students. Much of the project work took place outside of class time. The remainder of the school year was devoted to addressing the material from the mathematics textbooks in use prior to the projects intervention. Teachers assumed a facilitative role when their students were working on projects, and were free to revert back to more traditional instructional stances in the regular curriculum.

In contrast, at East Collins High School (ECHS), a few mathematics teachers planned collaboratively with local industry for nearly a year in developing what they intended to be a *Standards*-like curriculum. The teachers at ECHS adopted a philosophy they believed paralleled the *Standards*. Only after developing mathematical and pedagogical visions based on the *Standards* did they begin their search for textbooks and supplemental materials that would be consistent with their philosophy.

Despite the very different geneses of *Standards*-like reform in these two sites, the outcomes were strikingly similar. Both programs were endorsing a changed pedagogical vision wherein the teacher would be a questioner and a facilitator of learning. Teachers in both sites took risks in moving away from a familiar pedagogy in which they made all the decisions and took all responsibility for directing learning in their classrooms. Teachers at both schools reported that the facilitative roles that were being explored by teachers in the early reform efforts began to spread out "around every corner" of their teaching practice and into all of their classes. In both sites, teachers also noted that students initially experienced difficulty in working collaboratively in groups and accepting greater responsibility for their own learning. Teachers observed that students found the mathematics more exciting than doing repetitive exercises. According to one teacher at DVHS,

> Since projects help students teach themselves and each other, the material learned becomes a part of the student, and this gives the student a stake in his own education. The fact that projects extend and apply the students' knowledge and require them to write clearly and explain their solution means that students remember what they learn and see new connections. (p. 54)

Students discovered there were advantages to working in groups. These changes in students' attitudes towards mathematics helped teachers to take bolder steps. The teachers at DVHS began to apply group strategies to their regular classwork. Teacher interview comments indicated that lectures were getting shorter, and group-study time longer.

Research indicates that teacher beliefs influence classroom practice (Romberg, 1988a; Thompson, 1992; Underhill, 1988), but the role of experimental practice in influencing beliefs is less clearly understood. Teachers at ECHS were attempting to commit to the beliefs conveyed in the *Standards* before they began to implement their program. Teachers at DVHS were willing to try a different pedagogy with their projects, and as they saw evidence of the success of the "add-on" projects, they became more willing to undertake more comprehensive change. In both schools, mathematics teachers noted improvement in students' attitudes and understanding of mathematics. At ECHS, the teachers' modified belief system was confirmed and strengthened; at DVHS, practice with the projects began to reshape teacher beliefs—which changed practice further.

PEDAGOGICAL VISION

Although researchers had difficulty learning about the mathematical vision held by those at a site, it was much easier to learn about a site's pedagogical vision. Researchers found that teachers engaged in the reform of mathematics teaching were eager to discuss the shifts in teaching practice. In terms of finding aspects of the *Standards* documents in place in these sites, it is fair to say that pedagogical features were far more salient than mathematical features.

Analysis of pedagogical data across cases revealed a set of commonly held pedagogical emphases, including—

- student-centered teaching practice;
- more problem solving and less drill-and-practice activity;
- making mathematics enjoyable, interesting, and fun for students;
- opportunities for students to communicate about mathematics; and
- sharing authority for learning with students as a way of promoting their confidence and success;

Researchers found that it was frequently difficult to distinguish between pedagogical and mathematical visions, especially with regard to the first four Standards. The Standards on mathematics as problem solving, reasoning, connections, and communication convey pedagogical as well as mathematical values. Teachers who view mathematics as communications, connections, reasoning, and problem solving often see these four processes of doing mathematics as defining a particular pedagogical stance in the classroom. We observed that teachers found these four Standards to constitute an argument on behalf of student-centered classrooms, in which the teacher's role is more facilitative than directive.

Teachers explained that their roles were in transition and they were adopting a more facilitative presence in their classrooms. It was unclear precisely how teachers interpreted "facilitative," and how and where in the *Standards* they might have come to see the emphasis on facilitating as a reformist stance. Teachers did recognize that changes in their role required a realignment of responsibility for learning mathematics in the classroom, with students being given an ever-increasing share of that responsibility. In spite of these shifts, teachers noted that they found it difficult to move away from a teaching practice in which the teacher was the sole authority for right answers and "correct" mathematical knowledge. In his cross-case analysis of the five secondary school sites, Johnson (1995) found that the programs characterized as "student centered" enabled students to do mathematics in unthreatening ways more than in teacher-centered classrooms. Students and teachers both reported improvements in students' attitudes toward mathematics in these classrooms.

CONTEXT IN WHICH REFORM OCCURS

Perhaps the most striking finding of the study, one that is supported by the data from the full set of sites, has to do with the alignment of the *Standards* interpretation chosen by the sites and the context in which the school, classroom, or district happened to be situated. Repeatedly, we were able to note that the aspect of *Standards*-based reform a site chose to work on first, or to promote, or to consider was, in some clear way, well suited to a salient characteristic of the site's context. A suburban secondary school in a moderately high socioeconomic region, with a highly educated parent community, entered its reforms with a strong and visible technology thrust, thereby gaining support from the community for a longer term effort. East Collins focused on a real-world problem approach to curriculum that was well suited to the broad goals and commitment of their industrial partners. Desert View was strategic in embarking on a secondary school adaptation of a visible and well-received program at a local university, thereby establishing productive interaction and possibly credibility within their own community for their efforts. There are a number of other examples; the point is that for whatever reasons, given that the sites clearly chose different threads of the *Standards* as their initial emphasis, the better those choices seemed to fit with other trends in the community or the context, the further the efforts seemed to go. We have labeled this fit between contextural features and the reform emphasis chosen *congruence*.

There are a variety of contextual factors that seem critical in shaping efforts to improve mathematics teaching and learning. For example, Johnson (1995) used 29 coding categories to describe the different contextual features in his analysis of data from five of the sites observed in the R^3M project. Findings from cross-site data analyses indicated several factors that were instrumental in the initiation and support of reform efforts in mathematics. They include teachers' perceptions of mathematics reform, collaborations, and support for reform in the school community.

Teacher's Dispositions Toward Mathematics Reform

Reflection through collaboration can become an important link in the constructive process (Grouws & Schultz, 1996; Lieberman, 1995) and allows teachers opportunities "to look back on their teaching strategies, to reflect on the outcome of their behaviors, and to learn from experience" (Hart et al., 1991,

p. 6). Such experiences allow teachers to interpret the pedagogy they employ
and the ways in which the need they perceive for a changed pedagogy might be
translated into classroom practice. Reform is further complicated by the levels
at which individual teachers decide it is most appropriate for them to contribute
to those reform efforts (Grouws & Schultz, 1996; Lieberman, 1995). Oja and
Smulyan (1989), for example, found that stages of adult development relate to
individual motivation to participate in reform efforts, and they emphasized the
importance of understanding and allowing for these differences among teachers.
Beliefs that foster a disposition for change will come in varying degrees for
different teachers. Prawat (1992) suggested several impediments for teachers
trying to adopt a constructivist view of teaching and learning. One of these
impediments is a view of the learner and the curriculum as dichotomous instead
of interactive. This may explain why some teachers change seating
arrangements of students into groups but still dispense mathematics information
as though all students learned at the same rate and in the same way.

Teachers observed in our study varied in the degree they felt disposed and
motivated to engage in the mathematics reform effort. In this study, some
teachers' dispositions toward change allowed them to alter their pedagogy in a
number of new ways. Other teachers approached the changes in an incremental
and slower way.

Our analysis suggested that confidence was strongly connected to teachers'
tendency to reform their teaching practice. If administrators (or in the case of
Desert View, university mathematicians) helped teachers feel that they were the
mathematical professionals in a school, then they seemed to be more motivated
to find ways of improving pedagogical practice. This reflects the renegotiation
of authority that the *Standards* advocate in the classroom between student and
teacher, only at the level of teacher and administrator. Allowing teachers a
greater voice in their learning and their contributions to the decision-making
process at the school and district level seemed to facilitate efforts at change in
mathematics education.

For example, a teacher in one of the high schools in the study felt that staff
development needs were greater than the need for new textbooks. School
district policies, however, mandated that a fixed percentage of money be spent
on textbook purchases. Policies such as this, made at a level far removed from
the classroom, may impede the process of improvement in mathematics
teaching and learning. There were examples in other sites of cases in which the
teachers were being given more authority to influence policy decisions that
would affect their classroom practice.

Collaborative Activities

Nearly all teachers who participated in the project indicated that the most
important contextual factor was time to collaborate with others—especially
their teacher colleagues. Partnerships with universities and industry proved to
be important collaborations outside of the schools. Often, external funding

provided much of the impetus for these collaborations to develop. Collaboration was also key in determining the role of specialists and leaders in these sites.

Collaboration outside and inside. Collaboration allowed teachers to construct their own consensus about what they would value and pursue in their efforts to improve mathematics teaching and learning—their own interpretation of the reform. Perhaps the most striking commonality among the sites was the visibility of close, collegial communities of practitioners engaged in inquiry of various sorts. Teachers and staff in the sites invented various ways of enabling interaction. We saw teacher-initiated study groups, common planning time used to discuss lessons, specialists orchestrating meetings, regular after-school and before-school informal discussions, and sharing of articles and ideas "through the mailboxes." It seems that these structures developed relatively naturally and became entrenched, after time, in many of the sites. This phenomenon contributed to the sense of ongoing change, or a constant climate of reform, that has been noted by reform researchers as critical (Fullan, 1991). The data also suggest that teachers felt a need for frequent opportunities to collaborate with each other as they engaged in these shifts in practice. These findings parallel what Hart, Schultz, Najee-Ullah, and Nash (1991) found in the Atlanta Math Project.

Inside the schools, a close bond among teachers was observed in all of the sites represented by the case studies in this monograph and in most of the other 13 sites of the R³M study. Collaboration among colleagues was found by the researchers to be the most valuable (and most often mentioned) influence on teachers as they engaged in new pedagogical practice. At both East Collins and Desert View, for example, a collaboration based on friendship, humor, and trust led to a friendly esprit de corps, resulting in a noncompetitive, safe environment in which teachers were able to share ideas, concerns, and failures, and take risks. On the other hand, the Parker Springs case study illustrated that a safe, collaborative environment may not always be provided in a teacher's own school. And even in sites where the environment seemed collaborative and supportive, teachers recognized that this atmosphere was hard won.

> We're comfortable with one another as people and as teachers. But there again, you know, I think we have just such a wonderful opportunity here because we've been together for so long, we know one another. (Teacher, East Collins)

> So it was the chemistry at the time. The department was very collegial—got along very well and had for quite some time. And I don't think there were any real principalities there that you had to knock down in order to try something new .... I think, more than ever right now that the math faculty works together toward mathematical goals .... I think working on the projects really made them a lot closer as a faculty. (Assistant Principal, Desert View)

External collaborations or partnerships were also important and existed in three of the four case studies presented in this monograph. Partnerships with industry were in place both at Parker Springs and East Collins, and a university collaboration existed at Desert View. Industry and university partnerships also were established at other sites that participated in the R³M project.

Joan Ferrini-Mundy and Loren Johnson 119

Partnerships were often instigated or supported by special funding, and each partnership varied in its scope, functioning, and duration. Parker Springs is an example of a large district where a partnership with industry led to participation in the funded specialists program. East Collins teachers benefited from an industry partnership that lasted for 3 years, allowing for many of their mathematics reform efforts to become well established. The impetus from that collaboration was sustained by the school district's continued support of the evolving mathematics program. A university partnership was instrumental in assisting teachers at Desert View as they developed mathematics projects that became a significant part of change efforts taking place in their mathematics program.

Specialists and leaders of many sorts. At all of the elementary sites, and at many of the middle school and secondary sites, we found that a critical element in the reform effort was the presence of some version of a mathematics specialist or identified leader, although levels of official and recognized status within the district varied. This is not an argument for specialists but rather a statement of the reality that these sites included specialists and that for these sites, this seemed to be an important feature that worked in a positive way as a means of focusing energy and attention on issues of mathematics teaching and learning. Specialists came to be in these districts through a variety of routes: districts created new positions, outside grant monies allowed the creation of new positions, and classroom teachers recognized a need and assumed the specialist role in an unofficial capacity. The specialists' roles and the scope of their responsibilities varied considerably. In all cases, they were key to the reform efforts in the school and they helped sustain a presence for considering mathematics issues. They helped spread ideas, facilitate communications among teachers, plan and initiate staff development, and address political problems with administrators and community members.

Collaboration among teachers was formally encouraged through the Mathematics Specialist program in the Parker Springs School District. The specialist program was intended to provide in-depth in-service training and materials for selected K–3 teachers from each of the elementary schools of the district. Those teachers, in turn, served as agents for mathematics reform in their respective schools.

There were two levels of collaboration that evolved from the program. The first was the sharing of ideas between the specialist and other teachers in the schools. On a different level, the specialists themselves developed a collaborative enterprise whereby they supported one another to reach larger goals than could be reached individually. The researchers reported that

> the original 10 mathematics specialists talk about a "special" support system that has developed among them over the past 3 years. They say they are there to congratulate each other on their successes, console each other on their perceived blunders, to share their ideas, and to collaborate.

Support from the school community

Systematic community involvement in planning and sustaining attention to mathematics was evident in a significant way at several of our sites. This involvement ranged from the usual parent nights and notes of explanation sent home with work that didn't look like typical mathematics to formal committees, including parents, charged with determining goals for the mathematics program. The intricacies of dealing with parents and community members are substantial, and sites are learning a great deal about how critical such interactions are for sustaining and supporting change in mathematics teaching and learning. In some of the sites we studied, the parents became coinvestigators, in a sense, with the teachers. In one school district, for example, parents, businesses, and community leaders served with teachers for a period of 29 months to plan a reformed K–12 mathematics program.

INFLUENCE ON STUDENTS

Findings from the R^3M project indicate that programs that were student-centered provided new opportunities for students to do mathematics in nonthreatening ways. As teachers relinquished to students responsibility for learning, students and teachers both reported improvements in students' attitudes toward mathematics. These findings coincide with those of Wiske et al. (1992) and Tinto (1990), who found that negotiation of classroom authority became a feature in the reformed classrooms.

Equity Issues

The *Standards* documents clearly advocate *all* students having the opportunity to acquire mathematical power (NCTM, 1989), and they provide many suggestions for changing mathematical content and classroom practice to better serve all students. In the R^3M sites, there was evidence that changed pedagogical practices were beginning to affect students of all achievement levels. Many of the teachers we interviewed expressed their strong commitment to a higher quality, more relevant mathematics program for students traditionally thought of as being at risk. One teacher at Desert View summed it up as follows:

> I think our whole department is proud of the fact that we've had these kids who have never had any success in math—I mean they have flunked course after course after course—and we have intentionally looked for ways to give them some kind of success in math.

Equity issues were indicated as important either by teachers or mathematics supervisors in all of the R^3M high school sites (Johnson, 1995). Secada (1992) viewed this nation's present system of mathematics instruction as operating to perpetuate and broaden the inequities that exist for the "poor and ethnic minorities." According to Secada, reform efforts in curriculum and instruction "should first become effective with these students, and then be applied to other populations" (p. 654). The pattern in four of the five high schools in our study was to first establish the reform effort for those students who were not considered to be at risk. Then, if successful with these students, the reform efforts would be extended to include those students considered at risk (i.e., the poor and ethnic minorities). In the fifth site, an urban high school, the mathematics supervisor saw elitism in mathematics instruction as the single most important obstacle to mathematics reform. Reform efforts in that school were first addressing the concerns of the poor and ethnic minority students. The study does not include data that would help to illuminate the long-range implications of such different approaches.

Higher Expectations for Students

Teachers in several schools included in the study indicated that their expectations for student learning had been raised as a result of changed pedagogical practice. Teachers at Desert View were pleased at how much they could learn about student thinking from student work on mathematics projects.

Teachers in several secondary school sites reported that increased expectation of writing in mathematics was a major influence on students. Less clear is the influence of writing on mathematical understanding. Teachers observed that requiring frequent written explanations of problem-solving strategies was leading to improved writing skills. Early concerns about students' ability to write correctly later gave way to a sense of pride in students' accomplishments at several of the schools. There was also evidence of writing in mathematics classes at the elementary level. Deep Brook teachers, for example, indicated that they were using writing as a means of connecting and integrating mathematics with other disciplines.

Reduction in General Mathematics, Increased Persistence

Four of the five high schools in the R^3M project reported a trend of fewer and fewer students taking general mathematics courses. East Collins indicated that nearly all of its ninth graders were either taking Algebra I or Algebra II (geometry follows Algebra II), and approximately 91% of students taking Algebra I went on to take Algebra II. Green Hills was eliminating its general mathematics courses in favor of a focus on algebra and geometry through its Integrated Math 1, 2, and 3. Desert View indicated dramatic decreases in its general mathematics offerings (from 12 to 3 sections in the past three years), and more students were taking algebra and other higher level classes. Scottsville expected that calculus for all its students would be a reality by the year 2000. Pinewood's participation in the Equity 2000 program was an attempt to pull its students to higher levels of mathematics and possibly meet its goal of an algebra and geometry course for all its students. In most cases, these changes were new, and long-term achievement trend data were not yet available.

EVOLUTION OF REFORM EFFORTS

The way in which reform efforts in mathematics education evolved varied from school to school among the 17 sites. Catalysts for reform were not always the same from site to site. At some sites, varying degrees of curricular planning took place. At others, there were no specific plans in place for changing the mathematics curriculum. Nearly always, the support of administrators and other staff members was critical to the reform efforts. Questions about how long change takes were meaningless without a deep understanding of the context at a given site.

The Role of Curricular Planning

During the analysis of data, we were interested in determining what impact curricular planning had on reform efforts. Was it important that teachers in a school site had a plan to follow before they began to change instructional practice? Fullan (1993) warned that change is complex and unpredictable and wrote that "change is a journey and not a blueprint" (p. 21). He further suggested that "vision and strategic planning should come later" (p. 21), that too-careful planning can actually "blind" attempts at change.

We found that reform efforts were occurring at many schools in this study without reliance on formalized mathematics frameworks. This was certainly the case at DVHS, where the development of mathematics projects preceded any formalized curricular planning. A reformed mathematics curriculum was just being established at DVHS after 3 years of developing mathematics projects. At another of the R^3M sites, Johnson (1995) found that teachers initially resisted the newly developed district mathematics framework because the framework had little teacher input and ownership.

Several of the R^3M schools did long-term planning that influenced the materials, content, and pedagogy of the reform effort in their schools. ECHS teachers, for example, spent 9 months planning the kind of mathematics program they wanted for their students. During that time, teachers developed a new philosophy, organized new curricula, and began their search for what they believed to be the most appropriate materials to implement reform. Green Hills High School spent 29 months organizing and planning reform efforts.

Most often, reform efforts took place over time but not necessarily in a linear fashion. These efforts at improving mathematics education were in many senses improvisational and evolving. For example, Deep Brook administrators encouraged teachers to take risks in their teaching and experiment with new techniques and materials as their reform efforts evolved. A strategic plan that would "result in teachers using more innovative approaches to teach mathematics" was developed much later in their reform efforts. Even then, their planning was focused on providing ongoing professional development opportunities for teachers.

In theory, both extensive preplanning and taking a wait-and-see position on planning are advocated. Loucks-Horsley and Hergert (1985) suggested a seven-step linear process whereby reform efforts are planned, implemented, and modified. Much preplanning is required in their model, especially during the stages of assessment, goal setting, and identifying an ideal solution. This is in contrast to Fullan's (1993) position that strategic planning should occur much later in the process of reform. The data from the R^3M project indicate that preplanning becomes important to reform efforts in some sites; in other sites, reform efforts often take place in more evolutionary ways.

Administrative Support

The researchers observed that in several sites administrators and guidance staff were critical in acting as buffers to early criticisms about mathematics program changes. Criticisms from parents and others in the community were deflected in order to enable teachers time to work through some the difficulties encountered during their beginning ventures. For example, admonitions from East Collins administrators and guidance staff to parents to "give the teachers a little more time" really helped pave the way for public acceptance and support of the program.

Length of Time Involved

It is difficult to draw conclusions about the time required for shifts in mathematics teaching practice to take hold or about the quality of change taking place. The analyses of data from the five high schools in the study suggested that at two of the sites, efforts were well underway by the end of the third year. Teachers and other key players in those two sites perceived the first year of change efforts as being very worthwhile. Those early successes encouraged most of the other mathematics teachers to participate more actively during the following 2 years. Is a similar time frame what other schools should expect in their efforts at change? What about schools where mathematics teachers are struggling with many difficult issues and every new step taken seems monumental? Or where is there lack of consensus among teachers? Will the change process take longer in such sites?

The decisions by local teachers about which students get served first in efforts to reform may also have an impact on the time it takes a mathematics staff to bring about the reform they envision. Pinewood teachers' primary reform efforts sought to address the problem of elitism in mathematics courses by first providing quality learning opportunities in mathematics to those students characterized as being at risk. The principles that undergird any specific effort at improving mathematics education will be related in complex ways to the time involved in effecting the improvement.

Staff development was another key contextual factor, also related to the time it takes for change to take hold. The data indicated that certain kinds of staff development activities, including attendance at mathematics education conferences and release time for peer collaboration, were perceived by teachers as especially useful. These findings are corroborated by research by McLaughlin (1991), who reported that staff development activities that are decontextualized or disembodied from the daily realities of teaching (e.g., bringing in outside speakers unfamiliar with the school setting) have only limited value for teachers. The length of time required for reform efforts to occur will certainly be influenced by the staff development activities in a school, whether those activities are useful to teachers, and how well the activities align with the local context. There are, of course, many other contextual factors that influence how much time may be required to bring about a vision of reform (Fullan, 1993; Sarason, 1991).

IMPLICATIONS OF THE STUDY

Local context determined the form and nature of the mathematics reform in each of the sites in the study, but there were commonalities observed among some of the sites. We briefly discuss implications for instructional practice, implications for administrators and policy makers, implications for further research, and implications for *Standards*-based reform.

Implications for Instructional Practice

What are the implications for mathematics teachers who are interested in teaching different mathematics content and trying different pedagogical practice? Do the findings of this study provide direction for these teachers?

The ways in which teachers' beliefs and practices about mathematics learning and teaching change will be a highly individualized process. Our work indicates that there are many complex factors at both the individual and institutional levels that have an impact on teachers' needs and desires to enhance their understanding of how students learn mathematics and what pedagogy best suits that understanding. Adapting their practice is not always easy. The R^3M study portrays some of the struggles teachers faced in bringing about change in the way they believed that students learn and in the way they taught students.

Collaborating with peers, attending professional meetings of mathematics teachers, actively participating in curriculum development at the school and district levels, participating in partnerships with others outside the school, and being willing to take risks in changing the way they teach were ways that teachers in this study were bringing about reform in mathematics practice. Sometimes the struggle became intense as familiar practices were relinquished in favor of different models. Reflecting with colleagues on their shifting practice seemed critical for most teachers.

Implications for Administrators and Policy Makers

This study indicates that shifts in mathematics teaching and learning can be supported by innovative thinking and action by administrators and policy makers. For example, finding adequate resources for appropriate professional development in mathematics can be an important influence. Recognizing that mathematics teachers in the schools are the experts with respect to mathematics education will encourage more teacher input in curriculum development and instructional strategies. With this recognition of professionalism comes an obligation for policy makers to value the input of teachers in policy decisions.

Implications for Further Research

Johnson's analysis (1995) of the secondary school data in the R^3M project produced results consistent with the findings of studies done at the elementary school level (Ball, 1990; Ball, 1992b; Cohen, 1990; Lampert, 1990; Lappan & Theule-Lubienski, 1992; Silver, 1994; Wood, Cobb, Yackel, & Dillon, 1993). There are a number of factors that make teaching at the high school level different from teaching Grades K–4 and 5–8. Such factors include the specialization of teachers in one area of study; the success that high school teachers had in learning mathematics; the long-established tradition of a pedagogy dominated by directed teaching in mathematics at the secondary level; and the different structure of the high school as compared to the elementary school. Certainly, an extension of the high school study to include cross-case analyses for Grades K–8 data from the R^3M project would provide insights into the differences in how change occurs at the different grade levels.

Different types of research may be required to determine and describe some aspects of the reform efforts in mathematics education. Collaborative action research (Elliott, 1991; Oja & Smulyan, 1989; Oja, 1991) and studies about teaching (Ferrini-Mundy, 1994; Hart, 1991) have the potential to inform this area. Extended studies over longer time frames may also be warranted and would have the potential for addressing more adequately a number of questions raised by the R^3M project.

More research needs to be done that relates reformist interventions to student learning. The R^3M study was not designed to determine whether students were learning and understanding more mathematics as a result of their participation in *Standards*-like mathematics programs. Authentic assessment instruments were nonexistent in some of the sites visited and received only limited use in others. Most sites had been engaged in their reform efforts for 3 years or less, which further complicated the issue of assessment. More documentary research to gauge students' understanding of mathematics as outlined in the *Assessment Standards for School Mathematics* (NCTM, 1995) is certainly needed, together with description of practice.

Lieberman (1995) and Hart et al. (1991) have looked at how teachers acquire new knowledge and how this influences teaching practice. Additional transformative research is needed at the secondary level. A valuable contribution to existing research would be an in-depth study of how teachers' views of what is important mathematically change through the process of reflective collaboration with colleagues.

This study demonstrated the role of support mechanisms enabling teachers to engage in reflection on their practice. Schoenfeld (1992) identified the need for research at the systemic level to address the question "What changes in school and district structures are likely to provide teachers with the support they need to make the desired changes in the classroom?" (p. 365). Such research can inform reform efforts in mathematics education, but it also has implications for other disciplines.

Hart et al. (1991) raised two questions that have possible implications for further research:

- How do teachers' beliefs about learning mathematics, about teaching mathematics, and about the classroom environment change over time?
- How do teachers' beliefs about mathematical tasks and content change over time?

Implications for the Future of Mathematics Standards and for Looking at Standards-Based Reform

NCTM did pioneering work in launching the standards movement and has provided by example a working definition of "content standards" that has influenced standards in other areas. Much credit is due NCTM for this work. Yet in the 7 years since *Curriculum and Evaluation Standards for School Mathematics* was published, the landscape has changed considerably. Currently, the climate is characterized as a "national standards debate" (Kirst & Bird, 1996). This debate occurs at several levels. In the national policy arena, the debate seems to center on who will develop standards and whose authority will dominate the educational reform scene. In their 1996 summit, the nation's governors agreed to the development of state-level standards, for example. NCTM continues its commitment to providing leadership through the professional organization with its plans to revise and consolidate the current *Standards* documents. Recent discussion of the TIMSS research data contends that the standards provided by professional organizations are but one voice in a "tower of Babel" competing for influence in the reform of school mathematics.

In the world of curriculum, commercial textbook publishers' immediate response to the *Standards* seemed to be revision of tables of contents to align with the Standards topics, followed by a number of apparently more substantive efforts to incorporate into K–12 text series the topic areas advocated in the *Standards*. Determining the alignment of curriculum materials with Standards is a challenging problem worthy of continued examination.

At the state policy level, frameworks that reflect the NCTM *Standards* in some sense have been developed in most states. Often, the rather large-grain statements of the *Standards* have been particularized in state frameworks through translation into benchmarks or performance standards. Some researchers (St. John, 1996) argue that this modification through frameworks actually results in a dilution or distortion of the original *Standards* and even go so far as to conjecture that in states where the frameworks are looser, the *Standards* may stand a better chance of making their way directly into classrooms through teachers. The R^3M study suggests that both the systematic development of frameworks (as in some of our sites) and the more grassroots, teacher-driven development of *Standards*-like practice can lead to differences in classrooms.

A significant problem for the R³M project was the matter of actually determining the level of "alignment" of the classroom practice or school or district policy that we saw with aspects of the *Standards* documents. Our philosophy and stance in this descriptive study are addressed elsewhere in this document, but the problem of measuring *Standards*-based practice still remains. In their discussion of the politics of standards, Kirst and Bird (1996) concluded with a list of eight conflicts, including "If you approve standards that are too general ... you will be criticized that there is insufficient instructional guidance for teachers ... If you do approve pedagogy or detailed standards, you will be criticized because standards are too long, [and] complex, and overly control local practice." This observation might have given rise to a ninth conflict: If you approve standards that are too general, then determining whether they have been implemented in practice will be problematic. Ball (1992a) and Apple (1992) argued convincingly that the NCTM *Curriculum and Evaluation Standards* are underdetermined and that at best they will serve as a guide. These outlooks create a challenge in studying *Standards*-based reform, because observers' definitions of a *Standards*-based classroom can vary significantly. These differences were particularly striking in terms of what teachers at a given site would define as a *Standards*-like pedagogy. Sites varied widely in terms of the elements of the *Standards* message that became their central focus.

The *Standards* documents are clearly offered as a framework for discussion rather than a blueprint for change. They are intended to produce multiple interpretations and to invite discussion. At the same time, however, professionals in mathematics education certainly do hold strong opinions about what constitutes *Standards*-based mathematics teaching and learning. The matter of finding ways to judge, evaluate, and decide whether lessons, classrooms, teaching episodes, curricular materials, and school mathematics programs are truly *Standards*-based is fundamentally important and necessary yet also paradoxical. The R³M project stance is that providing descriptions can contribute to resolving these issues by promoting discussion and consensus in the field about what constitutes *Standards*-based practice.

Current and future efforts at Standards development and revision face questions that differed significantly from those faced in the development of the first version of the NCTM curriculum *Standards*. In particular, there is the matter of "grain size." McLeod et al. (1996) quoted Glenda Lappan's comments: "This [a goal of 15 statements per level] was important because it separated, from the very beginning, this document from anything that could be called scope and sequence ... This was a document about the big ideas" (p. 26). This consciously determined characteristic of the *Standards* is precisely what makes measurement and assessment of implementation difficult. The problem has actually been addressed a bit more explicitly in the *National Science Education Standards* (NRC, 1996), which states, "For the vision of science education described in the *Standards* to be attained, the Standards contained in all six chapters need to be implemented" (p. 3). The language in the introduction of the *Curriculum and Evaluation Standards for School Mathematics* is less directive: "Create a set of standards to guide the revision of the school mathematics curriculum..." (p. 1). "The standards should be viewed as facilitators of reform" (p. 2).

The challenging questions raised in determining the methodology and perspective for the R^3M project need to be considered in the revision of the *Standards* and possibly in the standards movement in general. What factors in the design of standards might lead to different options for evaluating their impact? And secondly, the finding that standards interpretation was arguably tied to context in several of the R^3M sites suggests that there are reasons for strong support of varied interpretations. Such contextually grounded interpretive initiatives may well stand the best chance of taking hold and making a difference in our students' mathematics knowledge.

References

Adelman, C., Jenkins, D., & Kemmis, S. (1976). Re-thinking case study: Notes from the Second Cambridge Conference. *Cambridge Journal of Education, 6*(3), 139–150.

Agar, M. (1980). *The professional stranger: An informal introduction to ethnography.* New York: Academic Press.

Altheide, D. L., & Johnson, J. M. (1994). Criteria for assessing interpretive validity in qualitative research. In N. K. Denzin & Y. S. Lincoln (Eds.), *Handbook of qualitative research* (pp. 485–499). Thousand Oaks, CA: Sage.

Apple, M. (1992). Do the *Standards* go far enough? Power, policy, and practice in mathematics education. *Journal for Research in Mathematics Education, 23*, 412–431.

Ball, D. (1990). Reflections and deflections of policy: The case of Carol Turner. *Educational Evaluation and Policy Analysis, 12*(3), 247–259.

Ball, D. (1992a). Implementing the NCTM Standards: Hopes and hurdles. In C. Firestone & C. Clark (Eds.), *Telecommunications as a tool for educational reform: Implementing the NCTM Standards* (pp. 33–49). Washington, DC: The Aspen Institute.

Ball, D. (1992b). Magical hopes: Manipulatives and the reform of math education. *American Educator, 16*(2), 14–18, 46–47.

Boyer, E. (1990). Reflections on the new reform in mathematics education. *School Science and Mathematics, 90*(6), 561–566.

California Mathematics Leadership Program. (1993). *Principles and caveats for comprehensive restructuring of mathematics education.* Unpublished paper.

Carpenter, T., & Moser, J. (1983). The acquisition of addition and subtraction concepts. In R. Lesh & M. Landau (Eds.), *The Acquisition of Mathematical Concepts and Processes* (pp 7–14). New York: Academic Press.

Clifford, J. (1986). Introduction: Partial truths. In J. Clifford & G. E. Marcus (Eds.), *Writing culture: The poetics and politics of ethnography* (pp. 1–26). Berkeley: University of California Press.

Cohen, D. (1990). A revolution in one classroom: The case of Mrs. Oublier. *Educational Evaluation and Policy Analysis, 12*(3), 311–329.

Cohen, D., & Ball, D. (1990a). Policy and practice: An overview. *Educational Evaluation and Policy Analysis, 12*(3), 233–239.

Cohen, D., & Ball, D. (1990b). Relations between policy and practice: A commentary. *Educational Evaluation and Policy Analysis, 12*(3), 331–338.

Davis, R. (1964). *Madison project: Discovery in mathematics.* Reading, MA: Addison-Wesley.

Denzin, N. K. (1978). *The research act: A theoretical introduction to sociological methods* (2nd ed.). New York: McGraw-Hill.

Dossey, J. (1992). The nature of mathematics: Its role and its influence. In D. Grouws (Ed.), *Handbook of research on mathematics teaching and learning* (pp. 39–48). New York: Macmillan.

Eisenhart, M. (1988). The ethnographic research tradition and mathematics education research. *Journal for Research in Mathematics Education, 19*, 99–114.

Eisenhart, M., & Howe, K. R. (1992). Validity in educational research. In M. LeCompte, W. Millroy, & J. Preissle (Eds.), *The handbook of qualitative research in education* (pp. 643–680). New York: Academic Press.

Elliott, J. (1991). *Action research for educational change.* Bristol, PA: Open University Press.

Erickson, F. (1986). Qualitative methods of research on teaching. In M. Wittrock (Ed.), *Handbook of research on teaching* (pp. 119–161). New York: Macmillan.

Ferrini-Mundy, J. (1992). *Recognizing and recording reform in mathematics education: Surveying and documenting the effects of the National Council of Teachers of Mathematics Curriculum and evaluation standards for school mathematics and Professional standards for teaching mathematics.* Original proposal submitted to Exxon Education Foundation.

Ferrini-Mundy, J. (1994, November). *Reform efforts in mathematics education: Reckoning with the realities.* Paper presented at the meeting "Reflecting on Our Work: NSF Mathematics Teacher Enhancement K–6," Arlington, VA.

Ferrini-Mundy, J., Graham, K., & Johnson, L. (1993, April). *Recognizing and recording reform in mathematics education: Focus on the NCTM Curriculum and evaluation standards for school mathematics and Professional standards for teaching mathematics.* Paper presented at the Annual Meeting of the American Educational Research Association, Atlanta, GA.

Ferrini-Mundy, J., & Johnson, L. (1993, October). *The place of problem solving in U.S. mathematics education K–12 reform: A preliminary glimpse.* Paper presented at the China-U.S.-Japan Joint Meeting on Mathematical Education, Weifang, China.

Ferrini-Mundy, J., & Johnson, L. (1994a). *Influence of NCTM's* Standards *on changed classroom practice: A preliminary look.* Unpublished manuscript, University of New Hampshire.

Ferrini-Mundy, J., & Johnson, L. (1994b). Recognizing and recording reform in mathematics: New questions, many answers. *Mathematics Teacher, 87*(3), 190–193.

Ferrini-Mundy, J., & Johnson, L. (1994c). *Standards in mathematics: Issues in documenting and monitoring progress.* Unpublished manuscript, University of New Hampshire.

Finn, C. (1993). What if those math standards are wrong? *Education Week, 23*(3), 36.

Fullan, M. (1991). *The new meaning of educational change.* New York: Teachers College Press.

Fullan, M. (1993). *Change forces: Probing the depths of educational reform.* New York: Falmer Press.

Gates, J. D. (1994). Report of the executive director. *Mathematics Teacher, 87*(6), 473.

Geertz, C. (1973). Thick description: Toward an interpretive theory of culture. In C. Geertz, *The interpretation of cultures.* New York: Basic books.

Goetz, J., & LeCompte, M. D. (1981). Ethnographic research and the problem of data reduction. *Anthropology and Education Quarterly, 12*, 15–70.

Goldsmith, L. T., & Schifter, D. (1993). Characteristics of models for the development of mathematics teaching. In J. Rossi Becker and B. Pence (Eds.), *Proceedings of the Fifteenth Annual Meeting of the North American Chapter of the International Group for the Psychology of Mathematics Education, Volume 2* (pp. 124–130). Pacific Grove, CA.

Graham, K., & Ferrini-Mundy, J. (1996, April). The mathematical "vision" of reform: Insights gained from the R³M project sites. In J. Ferrini-Mundy (Chair), *Documentive findings of reform efforts in mathematics education: Multiple perspectives from the R³M project.* Symposium conducted at the Annual Meeting of the American Educational Research Association, New York.

Grouws, D., & Schultz, K. (1996). Mathematics teacher education. In J. Sikula, T. Buttery, & E. Guyton (Eds.), *Handbook of research on teacher education.* New York: Simon & Schuster Macmillan.

Hart, L., Schultz, K., Najee-Ullah, D., & Nash, L. (1991). *Teacher change in the Atlanta Math Project: A process of implementing the NCTM Standards.* Unpublished manuscript.

Johnson, L. (1995). *Extending the National Council of Teachers of Mathematics' "Recognizing and Recording Reform in Mathematics Education" documentation project through cross-case analyses.* (Doctoral dissertation, University of New Hampshire, 1995). *Dissertation Abstracts International, 56*(05), 1696.

Johnson, L., & Ferrini-Mundy, J. (1995, September). Implications for curriculum development: Stories from the mathematics field. *NASSP Bulletin Quarterly Curriculum Issue, 25*, 1–4.

Kirst, M. W., & Bird, R. (1996). *The politics of developing and maintaining math and science standards.* Unpublished manuscript, Stanford University.

Koch, L. C., & Driscoll, M. (1996, April). Institutional change: Where does mathematics education fit in this context? In J. Ferrini-Mundy (Chair), *Documentive findings on reform efforts in mathematics education: Multiple perspectives from the R³M project.* Symposium conducted at the annual meeting of the American Educational Research Association, New York.

Kouba, V. (1989). Children's solution strategies for equivalent set multiplication and division word problems. *Journal for Research in Mathematics Education, 20*, 147–158.

Lampert, M. (1990). When the problem is not the question and the solution is not the answer: Mathematical knowing and teaching. *American Educational Research Journal, 27*(1), 29–63.

Lappan, G., & Theule-Lubienski, S. (1992). *Training teachers or educating professionals? What are the issues and how are they being resolved?* Paper presented at ICME-7 conference, Montreal, Canada.

LeCompte, M. D., & Preissle, J., with R. Tesch. (1993). *Ethnography and qualitative design in educational research.* New York: Academic Press.

Lieberman, A. (1995). Practices that support teacher development: Transforming conceptions of professional learning. *Phi Delta Kappan, 76*(8), 591–596.

Liggett, A. M., Glesne, C., Johnston, A., Hasazi, S., & Schattman, R. A. (1994). Teaming in qualitative research: Lessons learned. *International Journal of Qualitative Studies in Education, 7*(1), 77–88.

Lightfoot, S. L. (1983). *The good high school: Portraits of character and culture.* New York: Basic Books.

Loucks-Horsley, S., & Hergert, L. (1985). *An action guide to school improvement.* Andover, MA: The NETWORK.

Maxwell, J. A. (1992). Understanding and validity in qualitative research. *Harvard Educational Review, 62*(3), 279–300.

McGee-Brown, M. J. (1994, April). Ethics of interpretation. *Anthropology Newsletter.* American Anthropological Association.

McLaughlin, M. (1991). Enabling professional development: What have we learned? In A. Lieberman & L. Miller (Eds.), *Staff development for education in the '90s* (pp. 61–82). New York: Teachers College Press.

McLeod, D. B., Stake, R. E., Schappelle, B., Mellissinos, M., & Gierl, M. J. (1996). Setting the Standards: NCTM's role in the reform of mathematics education. In S. A. Raizen & E. D. Britton (Eds.), *Bold ventures: U.S. innovations in science and mathematics education. Vol. 3: Cases in mathematics education* (pp. 13–132). Dordrecht, The Netherlands: Kluwer.

Miles, M., & Huberman, A. M. (1994). *Qualitative data analysis* (2nd ed.). Thousand Oaks, CA: Sage.

Moon, J., & Hein, G. E., (1995) *Journeys to change.* Final Report of the Exxon Education Foundation's K–3 Math Specialist Program, program years 1988–94. Cambridge, MA: Lesley College, Center for Mathematics, Science, and Technology in Education and Program Evaluation and Research Group.

National Council of Teachers of Mathematics. (1989). *Curriculum and evaluation standards for school mathematics.* Reston, VA: Author.

National Council of Teachers of Mathematics. (1991). *Professional standards for teaching mathematics.* Reston, VA: Author.

National Council of Teachers of Mathematics. (1992, September). NCTM and Exxon join forces to study the impact of the *Standards. NCTM News Bulletin.*

National Council of Teachers of Mathematics. (1995). *Assessment standards for school mathematics.* Reston, VA: Author.

Oja, S. N. (1991). Adult development: Insights on staff development. In A. Lieberman & L. Miller (Eds.), *Staff development for education in the '90s* (pp. 37–60). New York: Teachers College.

Oja, S., & Smulyan, L. (1989). *Collaborative action research: A developmental approach.* New York: Falmer Press.

Parker, R. (1993). *Principles and caveats for comprehensive restructuring of mathematics education.* Unpublished manuscript.

Pelto, P. J., & Pelto, G. (1978). *Anthropological research: The structure of inquiry* (2nd ed.). New York: Cambridge University Press.

Peshkin, A. (1988). Understanding complexity: A gift of qualitative inquiry. *Anthropology and Education Quarterly, 19*(4), 416–424.

Phtiaka, H. (1994). What's in it for us? *International Journal of Qualitative Studies in Education, 7*(2), 155–164.

Porter, A. (1989). External standards for good teaching: The pros and cons of telling teachers what to do. *Educational Evaluation and Policy Analysis, 11*, 343–356.

Prawat, R. (1992). Teachers' beliefs about teaching and learning: A constructivist perspective. *American Journal of Education, 100*(3), 354–395.

Peterson, P. (1990). The California study of elementary mathematics. *Educational Evaluation and Policy Analysis, 12*(3), 241–245.

Punch, M. (1994). Politics and ethics in qualitative research. In N. K. Denzin & Y. S. Lincoln (Eds.), *Handbook of qualitative research* (pp. 83–97). Thousand Oaks, CA: Sage.

Research Advisory Committee of the NCTM. (1988). NCTM Curriculum and Evaluation Standards: Responses from the research community. *Journal for Research in Mathematics Education, 19*(4), 338–344.

Reys, R. (1992). The *Standards*? We did them last fall. *Arithmetic Teacher, 39*(5), 3.

Robitaille, D., Schmidt, W., Raizen, S., Mc Knight, C., Britton, E., & Nicol, C. (1993). *Curriculum frameworks for mathematics and science*. (Third International Mathematics and Science Study: Monograph No. 1). Vancouver, Canada: Pacific Educational Press.

Romberg, T. A. (1988a). *Changes in school mathematics: Curricular changes, instructional changes, and indicators of change* (Research Report Series No. RR-007). Center for Policy Research in Education.

Romberg, T. A. (1988b, April). *Policy implications of the three R's of mathematics education: Revolution, reform, and research*. Paper presented at the Annual Meeting of the American Educational Research Association, New Orleans.

Sarason, S. (1991). *The predictable failure of educational reform*. San Francisco: Jossey-Bass.

Schoen, H., Porter, A., & Gawronski, J. (1989). *Final report of the NCTM task force on monitoring the effects of the Standards*. Unpublished manuscript.

Schoenfeld, A. (1992). Learning to think mathematically: Problem solving, metacognition, and sense making in mathematics. In D. Grouws (Ed.), *Handbook of research on mathematics teaching and learning*. New York: Macmillan.

Schwandt, T. A. (1994). Constructivist, interpretivist approaches to human inquiry. In N. K. Denzin & Y. S. Lincoln (Eds.), *Handbook of qualitative research* (pp. 485–499). Thousand Oaks, CA: Sage.

Secada, W. (1992). Race, ethnicity, social class, language, and achievement in mathematics. In D. Grouws (Ed.), *Handbook of research on mathematics teaching and learning* (pp. 623–660). New York: Macmillan.

Shulman, L. (1986). Paradigms and research programs for the study of teaching. In M. Wittrock (Eds.), *Handbook of research on teaching* (pp. 3–36). New York: Macmillan.

Shulman, L. (1987). Knowledge and teaching: Foundations of the new reform. *Harvard Educational Review, 57*(1), 1–22.

Silver, E. (1990). Contributions of research to practice: Applying findings, methods, and perspectives. In T. Cooney (Eds.), *Teaching and learning mathematics in the 1990s* (pp. 1–11). Reston, VA: National Council of Teachers of Mathematics.

Silver, E. (1994, April). *Building capacity for mathematics instructional reform in urban middle schools: Context and challenges of the QUASAR project*. Paper presented at the Annual Meeting of the American Educational Research Association, New Orleans.

Spradley, J. (1979). *The ethnographic interview*. New York: Holt, Rinehart & Winston.

Spradley, J. (1980). *Participant observation*. New York: Holt, Rinehart & Winston.

St. John, M., Carroll, B., Oliver, M., Tambe, P., & Von Blum, R. (1996). *An evaluation of the NRC Partnership for Leadership: An interim report*. Unpublished manuscript.

Thompson, A. (1992). Teachers' beliefs and conceptions: A synthesis of the research. In D. Grouws (Eds.), *Handbook of research on mathematics teaching and learning* (pp. 127–146). New York: Macmillan.

Tinto, P. (1990). Students' views on learning proof in high school geometry: An analytic approach. (Doctoral dissertation, Syracuse University, 1990). *Dissertation Abstracts International, 51* (04), 1149.

Underhill, R. (1988). Focus on research into practice in diagnostic and prescriptive mathematics: Mathematics teachers' beliefs: Review and reflections. *Focus on Learning Problems in Mathematics, 10*(3), 43–58.

Weiss, I. (1992). The road to reform in mathematics education: How far have we traveled? (Pilot study prepared for the National Council of Teachers of Mathematics.) Chapel Hill, NC: Horizon Research.

Weiss, I. R., Matti, M. C., & Smith, P. S. (1994). *Report of the 1993 national survey of science and mathematics education*. Chapel Hill, NC: Horizon Research, Inc.

Winter, M. J., & Carlson, R. J. (1993). *Algebra experiments 1: Exploring linear functions*. Menlo Park, CA: Addison-Wesley.

Wiske, M., Levinson, C., Schlichtman, P., & Stroup, W. (1992). *Implementing the Standards of the National Council of Teachers of Mathematics in high school geometry*. Technical Report. Cambridge, MA: Harvard Graduate School of Education, Educational Technology Center.

Wolcott, H. (1982). Differing styles of on-site research, or "If it isn't ethnography, what is it?" *Review Journal of Philosophy and Social Science, 7*, 154–169.

Wolcott, H. (1984). *The man in the principal's office: An ethnography.* Prospect Heights, IL: Waveland Press.

Wolcott, H. (1988). "Problem finding" in qualitative research. In H. Trueba & C. Delgado-Gaitan (Eds.), *School and society: Learning content through culture* (pp. 11–35). New York: Praeger.

Wolcott, H. (1990a). On seeking—and rejecting—validity in qualitative research. In E. W. Eisner & A. Peshkin (Eds.), *Qualitative inquiry in education: The continuing debate* (pp. 121–152). New York: Teachers College Press.

Wolcott, H. (1990b). *Writing up qualitative research* (Sage University Press Series on Qualitative Research, Vol. 20). Newbury Park, CA: Sage.

Wolcott, H. (1992). Posturing in qualitative inquiry. In M. D. LeCompte, W. L. Millroy, & J. Preissle (Eds.), *The handbook of qualitative research in education* (pp. 3–52). New York: Academic Press.

Wolcott, H. (1994). *Transforming qualitative data: Description, analysis, and interpretation.* Thousand Oaks, CA: Sage.

Wood, T., Cobb, P., Yackel, E., & Dillon, D. (Eds.). (1993). *Rethinking elementary school mathematics: Insights and issues. Journal for Research in Mathematics Education* Monograph Series, Number 6. Reston, VA: National Council of Teachers of Mathematics.

Woods, P. (1986). *Inside schools: Ethnography in educational research.* London: Routledge.

Appendix A

Observation and Interview Guides

R^{3}M SITE-VISIT WRITE-UP COVER SHEET
(FORM Z)[1]

Site:_____

Date of site visit:_____

Documenters' names:_____

Date of submission of write-up:_____

How was the write-up done? (Who did which part of the write-up? How did you collaborate or divide up the work?) _____

_____

Date of site visit:_____

Summary of site visit schedule (please indicate how all time was spent there and with whom you spoke at each appointment, including the person's title/responsibility):_____

Pseudonyms (please use these hereafter):

    Actual name _____   Pseudonym _____

Please list everything that is included in this packet:

    Interview summaries for interviews with:_____

    Classroom observation summaries:_____

    Summary of meeting with teachers:_____

    Write-up:_____

    Other: _____

What materials did you collect beforehand or during the visit? Please list. Please enclose copies._____

_____

Do you feel that this site should be selected as one of six to be visited for an extended period in the spring, given what you understand of the R^{3}M goals? Why or why not?_____

---

[1]Form Z served as a reminder to documenters of the particular items that were to be collected during the site visit. It also served as record of the items to be stored at R^{3}M headquarters.

If your answer is yes—

A. What specific areas are interesting and should be pursued in detail in the next visit?_____

B. Who is a logical candidate for the "inside" documentation team member? Why? Will there be any political difficulties?_____

C. What self-documentation should we request from the site between now and a second visit (curriculum outlines, student work from particular teachers' classes, in-service schedules and materials, etc.)?_____

Logistics:

Were the people in the site prepared for the visit? What should have been done
differently in advance?_____

Who else did you wish you had talked with? Why? Could a telephone interview
help?_____

Were there any concerns about publicity? Had anything been released locally?
Please provide copies of any press clippings, etc. if possible._____

_____

Do you think we will have any problems or "awkwardnesses" with this site?
Should we be anticipating any problems?_____

Individual documenters should feel free to forward individual comments.

WRITE-UP QUESTIONS (FORM F)[2]

INSTRUCTIONS: Feel free to create your own form, but please respond to these questions. We suggest that you formulate an answer to the question, and then that you provide as much supporting evidence as possible. Supporting evidence can come from anywhere—documents, interviews, observations, etc. Be very specific as you do so. For example, say, "The principal said..."; "The mathematics department's mission statement says..."; "In her classroom, Teacher X did ... and the students did..."; "In the teachers' room, two teachers were talking and they said..." If you draw a conclusion based on your own impressions but cannot find clear documentation, please include, but make it clear that you have only an impression without clear support.

Note: These write-ups will be used only by R³M staff at this point. PLEASE DO NOT SHARE THIS WITH THE SITE AT THIS TIME; FOR SITES THAT ARE TO BE VISITED A SECOND TIME, WE MAY LATER DECIDE ABOUT SHARING THIS IN SOME WAY. FOR NOW, THE ONLY COPIES

_____

[2]The completed Form F served as the primary data summary for the documentation project. Later analyses of data across sites also followed the six areas cited on the following pages.

SHOULD BE IN YOUR PERSONAL FILES AND IN THE PROJECT FILES. Notice that each question calls for description and then interpretation.

1. DESCRIBE THE "MATHEMATICAL VISION" HELD BY THE PEOPLE IN THE SITE. (What are their goals for their mathematics program? What kinds of mathematics learning do they hope their students will experience? What features of the *Standards* emerge as they describe their mathematical vision? What do the people in the site see as worthwhile mathematical tasks or important mathematical ideas? Is there alignment among teachers and administrators concerning this vision? Does the mathematics program emphasize problem solving, communication, reasoning, conjecturing, and mathematical connections? Can the classrooms be described as mathematical communities? Why? IS THE MATHEMATICAL VISION THAT THEY HOLD BEING BROUGHT TO LIFE IN THE CLASSROOM? WHAT IS HAPPENING, MATHEMATICALLY, IN THIS SCHOOL? PROVIDE AS MUCH SPECIFIC EVIDENCE AS POSSIBLE.

2. DESCRIBE THE "PEDAGOGICAL VISION" HELD, RELATIVE TO MATHEMATICS, BY THE PEOPLE IN THE SITE. (What pedagogical philosophy is articulated in the site? How can you tell? What approaches, strategies, and ways of teaching mathematics are important here? What features of the *Standards* emerge as they describe their pedagogical vision? What do the people in the site believe to be effective pedagogical practices? Is there alignment among teachers and administrators concerning this vision?) IS THE PEDAGOGICAL VISION THAT THEY HOLD IN THIS SITE BEING BROUGHT TO LIFE IN THE CLASSROOM? WHAT IS HAPPENING, PEDAGOGICALLY, IN THIS SCHOOL? PROVIDE AS MUCH EVIDENCE AS POSSIBLE.

3. DESCRIBE HOW CONTEXTUAL FEATURES ARE INFLUENCING, BOTH POSITIVELY AND NEGATIVELY, THE TEACHERS' EFFORTS TO CHANGE THEIR MATHEMATICS PRACTICE. (What has happened with inservice training, outside consultants, outside funding, the school and district administration, and the community with regard to the teacher, the role of materials, scheduling and structural matters, and policies about evaluation and testing.)

4. DESCRIBE THE WAY THAT THE MATHEMATICAL AND PEDAGOGICAL PRACTICES IN THIS SCHOOL ARE AFFECTING STUDENTS. (Are students engaged? What kind of mathematics are they learning? Do the pedagogical approaches seem effective? What is the nature of the discourse and mathematical communication? Are students learning to reason and solve problems? Are they experiencing mathematical connections?)

5. DESCRIBE THE EVOLUTION OF THE MATHEMATICS PROGRAM IN THIS SCHOOL. (How has the program gotten to be the way it is? What factors are important? Where do people in the site feel there is a need for continued growth, support, and development?)

6. OTHER. (Comment on other important observations that will help people understand the way in which this site is interpreting and adapting the ideas of the NCTM *Standards*.)

Appendix B

A Third-Grade Math Episode, Seen Through the Lens of One Mathematician

Hyman Bass, *Columbia University*

This episode is extracted from the report of Laura Coffin Koch and Mark Driscoll, "Institutional Change: Where Does Mathematics Education Reform Fit in This Context," which contributed to the R³M Project designed to "Recognize and Record Reform in Mathematics Education," in particular, the impact of the NCTM *Standards*. It describes a classroom scene of the K–3 math specialist, Barbara Myers, at the Armstrong Elementary School (pseudonym).

THE EPISODE

Upon entering the room, as a preliminary to the day's lesson, Barbara Myers puts the following problem on the board:

> They gave Sadie a bath with flea-tick shampoo. Twenty-eight fleas were seen to wash down the drain. After the bath, how many fewer legs were crawling on Sadie?

BM: What do we need to find out?

[The students quickly realize that they need to know how many legs there are on a flea. One student proposes that because a flea is an insect, it has six legs. So the problem is recognized to be how many legs there are on 28 six-legged fleas.

After some time of student work, Sally offers her solution. She goes to the board and writes (in full) $6 + 6 + 6 + \ldots + 6$ (28 sixes, written in a column) and does the addition. After an addition error, which she corrects herself, she persists to the final answer, 168.]

BM: Any other solutions?

Colin: The problem is 6 multiplied by 28, so it's $28 + 28 + 28 + 28 + 28 + 28$, which he adds to get 168.

BM: Any others?

Debby: It's $6 \times 20 + 6 \times 8 = 120 + 48 = 168$.

[The episode then ends with a brief commentary by BM on the different approaches and the use of "short cuts."]

COMMENTARY

1. The problem involved an early exposure to multidigit multiplication, with children who presumably knew addition, and the multiplication table (single digit multiplication).

2. The problem was "ill posed," in that its statement required information not stated yet essential to give it a precise mathematical formulation. The students were called on to identify the missing information and to supply it from prior knowledge.

3. Once modeled as multiplication, 28 multiplied by 6, Sally's solution illustrated a direct assault, parallel with what is seen in the problem context, of treating this multiplication as iterated addition. It is correct, methodologically transparent, and labor intensive, and therefore error prone.

4. Colin invokes the commutativity of multiplication and makes astute use of it in this instance to greatly simplify the computation.

Note that the context visually presents 28 packets of six legs each, to be added, but there is nowhere visible in the context any six arrays of 28 legs each. Thus, the step taken by Colin is an illustration of some of the freedom of operation afforded by mathematical abstraction, once the problem is mathematically disengaged from its context.

5. Just as Colin's approach illustrates the commutative law, Debby's illustrates and makes deft use of the distributive law; 28 multiplied by 6 = (20 + 8) multiplied by 6 = (20 multiplied by 6) + (8 multiplied by 6).

The effect of this is to reduce two-digit multiplication to the multiplication table (one-digit multiplication) plus simple rules about how to do the bookkeeping with the zeroes. In fact, this method is completely generalizable, and Debby has discovered the genesis of our standard algorithm for multidigit multiplication, thus also exhibiting an important appreciation of the nature of place-value notation.

6. Barbara Myers' discussion of "short cuts" is an entree into the important issue of the cost and complexity of various problem-solving strategies, an issue worthy of more focused attention by students. Programs for computers, which embody frequently executed algorithms, have to pay serious attention to such cost considerations; this has led to a whole new field of theoretical computer science, complexity theory. This discussion emphasizes that one calculates with ideas as well as with algorithms, (which are themselves the products of ideas).

7. Is this a "real-world problem?" Well, not in any serious pragmatic sense. It has a sketchy "context," but it is pretty fanciful. But that is not an issue here. The aim is to, in a relatively unencumbered way, expose the children to an interesting mathematical task with enough context to make it convincing and for which the students did not yet have the standard mathematical tools. They were invited to invent them, or cope in any ways they could manage from prior knowledge and mathematical experience and intuition.

8. Is this a problem in computation? In basic skills? There is a lot of confusion and, in my view, misguided thinking in the public use of these descriptors, which now seem to evoke all the scorn heaped on past excessive educational practices. First of all, yes, this is a problem in computation. In fact, vast parts of mathematics can be understood as dealing with problems of computation. There

is the matter of executing computations with clearly designed algorithms. When the latter are available, they are a powerful resource, and there should be good understanding and practice in their use. But there is also the matter of discovering or inventing algorithms, of analyzing and evaluating them. More generally, mathematics involves in essential ways the ideas of computation, the use of ideas as tools of computation, and the concept that some problems have no algorithmic solution at all within certain frameworks (straight-edge and compass trisection of angles and solution of equations of degree greater than 4 by radicals, for instance). The question about basic computational skills is not whether or how little we should teach them or whether we should transfer this domain entirely to computer execution but rather how we should teach them; they are essential, and central to mathematical education. The episode above illustrates an enlightened approach to this question, which is to treat the very algorithmic process itself as an object of inquiry and student manipulation and not simply as an enshrined ritual or a "black box" that provides a useful deliverable.

These clarifications of meaning are crucial when one contemplates the prospect of "recognizing" and "measuring" progress toward the kinds of reform in mathematics education advocated in general ways by the *Standards*. Naive promotion of "real-world problems," and deprioritizing of "basic computational skills" can, when carelessly expressed, subvert some high quality mathematical teaching practice, which is integral to a proper, and ecumenical, reading of the *Standards*.

Recognizing and Recording Reform in Mathematics Education (R³M) Documentation Team

(affiliations during project work)

Joan Ferrini-Mundy, *Project Director*, University of New Hampshire

Clem Annice, Canberra University (Belconnen, Australia)

Gabrielle Brunner, Milton Academy (Milton, MA)

Mark Driscoll, Education Development Center

Julie Fisher, National Council of Teachers of Mathematics

Beverly Ferrucci, Keene State College

Karen Graham, University of New Hampshire

Loren Johnson, University of New Hampshire

Laura Coffin Koch, University of Minnesota

Diana Lambdin, Indiana University

Linda Levine, Orange County Public Schools (Orlando, FL)

Recognizing and Recording Reform in Mathematics Education (R³M) Advisory Board Members
(affiliations during project work)

Mary M. Lindquist, *Chair*
President, National Council of Teachers of Mathematics

Deborah Ball, Michigan State University

Joan Ferrini-Mundy, University of New Hampshire

James D. Gates, Executive Director, National Council
of Teachers of Mathematics

Marilyn Hala, National Council of Teachers of Mathematics

Linda Levine, Orange County Public Schools (Orlando, FL)

Edward Silver, University of Pittsburgh

Donald M. Stewart, The College Board

Lynn A. Steen, Mathematical Sciences Education Board

Robert Witte, Exxon Education Foundation